# Synthesis Lectures on Biomedical Engineering

This series consists of concise books on advanced and state-of-the-art topics that span the field of biomedical engineering. Each Lecture covers the fundamental principles in a unified manner, develops underlying concepts needed for sequential material, and progresses to more advanced topics and design. The authors selected to write the Lectures are leading experts on the subject who have extensive background in theory, application, and design. The series is designed to meet the demands of the 21st century technology and the rapid advancements in the all-encompassing field of biomedical engineering.

Abdelbary Elhissi · Dana Elkhalifa ·
Iftikhar Khan · Waqar Ahmed

# Proliposomes: A Manufacturing Technology of Liposomes for Pulmonary Drug Delivery

Abdelbary Elhissi
Department of Pharmaceutical Sciences
College of Pharmacy, QU Health
Qatar University
Doha, Qatar

Iftikhar Khan
Faculty of Science
School of Pharmacy and Biomolecular
Sciences
Liverpool John Moores University
Liverpool, UK

Dana Elkhalifa
Department of Pharmacy
Aspetar Orthopedic and Sports Medicine
Hospital
Doha, Qatar

Waqar Ahmed
School of Engineering and Physical Sciences,
College of Health and Science
University of Lincoln
Lincoln, UK

ISSN 1930-0328        ISSN 1930-0336  (electronic)
Synthesis Lectures on Biomedical Engineering
ISBN 978-3-319-01296-4        ISBN 978-3-319-01297-1  (eBook)
https://doi.org/10.1007/978-3-319-01297-1

This Springer imprint is published by the registered company Springer Nature Switzerland AG
The registered company address is: Gewerbestrasse 11, 6330 Cham, Switzerland

If disposing of this product, please recycle the paper.

# Preface

This monograph is made of five chapters and is directed to the scientific community in the fields of pulmonary drug delivery, liposomes, and generally the field of advanced drug delivery systems. Proliposome technologies represent a promising liposome manufacturing approach for liposomes, since they offer advantages over traditional liposome formulation methods, in terms of stability, convenience, and scalability. Below is a brief overview of each chapter.

Chapter 1 starts by giving a comprehensive overview of pulmonary drug delivery systems and inhalation devices. The chapter demonstrates the growing incidence of lung diseases such as asthma and chronic obstructive pulmonary disease (COPD), which remain leading causes of morbidity and mortality worldwide. The chapter reviews the advantages of pulmonary drug delivery as a non-invasive route for administration of therapeutic materials.

Furthermore, the historical context and current designs of devices used for pulmonary inhalation are thoroughly reviewed. Inhalation devices are categorized into three main types, which are pressurized metered dose inhalers (pMDIs), dry powder inhalers (DPIs), and medical nebulizers. These devices were subcategorized, in order to successfully elaborate, in the subsequent chapters, on their suitability for delivering liposomes and proliposomes. Some formulation characteristics influencing the efficiency of drug delivery from these devices were discussed.

For DPIs, formulation strategies employed to enhance drug delivery through these devices were discussed, emphasizing the role of carrier particles in improving powder flowability and particle dispersion during inhalation.

The design of pMDIs was also explained, detailing their evolution since the first pMDI device was introduced to the market in 1956. The transition from using chlorofluorocarbons (CFCs) to relying on hydrofluoroalkanes (HFAs) as propellants was justified by the environmental safety of the latter compared to the earlier.

Medical nebulizers are the third type of inhalation devices that are commonly used for delivering liquid medications (solutions or dispersions) to the lung. Nebulizers are commonly used for continuous aerosol generation over a period of time, usually to "dryness" (i.e. when aerosol generation is completely ceased). Nebulizers are subcategorized into air-jet, ultrasonic, and vibrating-mesh nebulizers, and details about their operational principles and impact of their design on aerosol performance are explained. Among the inhalation devices, no single device or mechanism of operation is regarded universal for all patients or conditions.

In Chap. 2, proliposome technologies are introduced with a special highlight on the formulation strategies of proliposome-based drug delivery systems, which are argued, with the support of research findings, to be superior alternatives to traditional liposome formulations, such as those made using the thin-film hydration method. Proliposomes, which convert into liposomes upon hydration, may offer enhanced stability, convenient handling, and improved scalability compared to the thin-film made liposomes.

In this chapter, liposomes are defined and described in terms of composition, morphology, and size. Liposomes are classified into multilamellar vesicles (MLVs), large unilamellar vesicles (LUVs), and small unilamellar vesicles (SUVs) and the preparation methods for each type of liposomes are described. The advantages and limitations of different liposome preparation methods are also outlined.

This chapter also discusses freeze-drying for extending liposome shelf life. Proliposome technologies are presented as a convenient alternative to overcome the instability challenges associated with manufacturing liposomes. Proliposomes can be categorized into particulate-based and solvent-based proliposomes, with the former involving solid carriers that can incorporate a drug-lipid mixture, and the latter employing alcohol to create lipid solutions/mixtures that form liposomes upon hydration. Proliposome technologies offer advantages such as improved formulation stability during storage, ease of handling, and higher drug entrapment efficiencies compared to traditional liposomes (e.g. liposomes made by thin-film hydration).

Furthermore, this chapter discusses the scaling up preparation methods of proliposomes, including fluidized-bed coating and spray drying. The discussion emphasizes the importance of optimizing various aspects of proliposome formulations, including carrier selection, lipid composition, and process parameters, to enhance drug encapsulation efficiency and drug delivery effectiveness.

Chapter 3, entitled "Formulation Approaches for Proliposomes in Pulmonary Drug Delivery", discusses the promising role of proliposomes as an inhalable drug delivery system for the treatment of lung diseases, and in some occasions, for treating illness in organs distant from the lung. Proliposomes can generate liposomes upon hydration, making them particularly advantageous for pulmonary drug delivery due to their good stability, high drug encapsulation efficiency, and ability to incorporate a wide range of therapeutic molecules. Careful design of proliposome formulation involves selection of the right excipients and preparation technique. This involves the selection of the right lipid composition, and carbohydrate carrier (if needed), in addition to excipients that may enhance formulation performance.

This chapter also comes across many characterization techniques necessary for assessing the physicochemical properties of proliposomal formulations and the resultant liposomes following hydration. These techniques include X-ray diffraction, particle size analysis, zeta potential determination, drug encapsulation estimation, and assessing the aerodynamic behavior of the formulations. The chapter also outlines various analytical techniques used for assessing aerosolization such as laser diffraction for analyzing droplet size distribution and inertial impaction and high performance liquid chromatography (HPLC) for the determination of drug in fine particle fraction (FPF).

The chapter then transitions to discussing formulation approaches for the different pulmonary delivery devices, including pressurized metered-dose inhalers (pMDIs), dry powder inhalers (DPIs), and medical nebulizers. The development of formulations for delivery via these devices is explored in details, emphasizing the importance of selecting the right excipients for delivery via the right device. Finally, the chapter concludes with an overview of the therapeutic efficacy and toxicity studies necessary for validating the safety and effectiveness of proliposome formulations *in vivo*. It emphasizes the need for ongoing research and innovation in order to enhance the delivery and clinical impact of proliposome-based formulations.

The therapeutic applications of inhalable proliposomes and liposomes generated from proliposomes are discussed in Chap. 4. Proliposomes can effectively deliver drugs directly to the respiratory tract, making them potentially suitable for the treatment of local lung conditions such as asthma, chronic obstructive pulmonary disease (COPD), cystic fibrosis, pulmonary infections, lung cancer, etc. Research studies indicated that proliposomes can significantly improve the stability and bioavailability of various medications, enabling targeted delivery with maximized therapeutic benefit and minimized systemic toxicity.

The liposomal amikacin (Arikayce®) has been approved by FDA, suggesting future developments may introduce equivalent proliposome formulations for clinical use. The document reviews ongoing studies displaying the promising use of proliposomes in treating respiratory tract infections, particularly tuberculosis (TB), where they can encapsulate first-line anti-tubercular medications like rifampicin and isoniazid. Several studies have demonstrated favorable aerosolization properties, sustained release, and improved efficacy of proliposomal formulations against mycobacterium tuberculosis. Furthermore, proliposomes have been highlighted as promising formulations for managing asthma by using water-soluble antiasthma drugs like salbutamol and insoluble drugs like the steroid beclometasone dipropionate. Moreover, the use of proliposomal formulations in the treatment of lung cancer (e.g. using paclitaxel) may improve efficacy. Importantly, the investigation of proliposome formulations in gene therapy for the treatment of lung diseases was also highlighted. Liposomes serve as non-viral vectors that can effectively deliver therapeutic genes to specific cells, with studies indicating their success in reducing tumor growth using experimental animals.

Chapter 5 provides the general conclusions of proliposomes in terms of formulation, and devices used for delivery of proliposomes and liposomes generated from proliposomes. Moreover, the chapter explains potential therapeutic applications of liposomes prepared using the proliposome technologies. Finally, the chapter proposes some future directions in the field of proliposomes for pulmonary drug delivery.

Doha, Qatar                                                         Abdelbary Elhissi
Doha, Qatar                                                             Dana Elkhalifa
Liverpool, UK                                                          Iftikhar Khan
Lincoln, UK                                                             Waqar Ahmed
2025

**Acknowledgements** We would like to thank Prof. Kevin M. G. Taylor, Emeritus Professor of Pharmaceutics at UCL School of Pharmacy, for his unwavering inspiration and leadership in the field of pulmonary drug delivery of liposomes. His publications and personal guidance have been invaluable to our learning journey and research in the field of liposomes and proliposome drug delivery systems.

We would like also to state that Prof. Abdelbary Elhissi and Ms. Dana Elkhalifa are co-first authors of this book, having contributed equally to its authorship.

The authors would like to express their sincere thanks Dr. Sakib Yousaf, Senior Lecturer at the School of Pharmacy and Biomolecular Sciences, Liverpool John Moores University, for his contribution to the authorship of Chap. 1 "Introduction: Pulmonary Drug Delivery and Inhalation Devices". Dr. Yousaf was involved in reviewing previous studies on different inhalation devices. The authors also extend their appreciation to Dr. Hassaan Anwer Rathore, Associate Professor at the College of Pharmacy, QU Health, Qatar University, for his contribution to the authorship of Chap. 4 "Therapeutic Applications of Proliposomes in Pulmonary Drug Delivery". Dr. Rathore reviewed relevant in vivo studies and contributed to the chapter's critical evaluation.

The authors would like to acknowledge that the topic of this monograph is connected to the grant QUCG-CPH-21/22-1, secured by the Lead PI, Prof. Abdelbary Elhissi. Chapter 4 specifically relates to the grant QUCG-CPH-21/22-1, awarded to Prof. Abdelbary Elhissi, as well as the grant QUCG-CPH-24/25-547, secured by Dr. Hassaan Anwer Rathore. Our sincere thanks to Qatar University for the generous financial support through these two collaborative grants.

**Competing Interests** The authors have no competing interests to declare that are relevant to the content of this manuscript.

# Contents

# Introduction: Pulmonary Drug Delivery and Inhalation Devices

**Abstract**

In this chapter, a comprehensive review of devices intended for use in pulmonary inhalation was introduced, from both historical context and technical perspective. Inhalation devices are of three main types, which are pressurized metered dose inhalers (pMDIs), dry powder inhalers (DPIs) and medical nebulizers. For DPIs, the role of carrier was discussed and the need for active inspiration of the patient was emphasized. For pMDIs, the significance of transition from the ozone-depleting chlorofluorocarbon propellants to the environmentally safer ones, namely hydrofluoroalkanes (HFAs), was reported. Medical nebulizers were discussed in terms of their subtypes and benefit using traditional solutions or dispersions of the medication. Recently, many researchers have considered soft mist inhalers (SMIs) as an additional category of inhalation devices, whilst others may consider it a nebulizer-like device taking into account it delivers liquid formulations; thus, we also reviewed SMIs.

## 1.1 Introduction

The terms respiratory or pulmonary are most often used interchangeably relating to the respiratory system or lungs as a whole. The pulmonary system primarily functions as an exchange interface for $CO_2$ and $O_2$ between the environment and human body. Tiny hairs, referred to as cilia, protect the respiratory tract as air passes through the nose/mouth, throat, trachea (connecting the larynx to the bronchial components), bronchioles, and finally reaching the air sacs (i.e., alveoli), where gas exchange takes place (Liang et al. 2015).

The rising popularity and sustained interest in the pulmonary route may be attributed to its non-invasive nature, in addition to offering a large-vascularized surface area of

© The Author(s), under exclusive license to Springer Nature Switzerland AG 2025
A. Elhissi et al., *Proliposomes: A Manufacturing Technology of Liposomes for Pulmonary Drug Delivery*, Synthesis Lectures on Biomedical Engineering,
https://doi.org/10.1007/978-3-319-01297-1_1

~ 100 m$^2$ (Khan et al. 2023; Malamatari et al. 2020). The use of the pulmonary system for inhalation of pharmacologically active materials may be traced back to as far as 4000 years ago, where there is evidence that the smoking of *Atropa belladonna* was used in the treatment of coughs (Grossman 1994). Additionally, in the early nineteenth and twentieth centuries, tobacco leaves were utilised through mixing with stramonium powder for the treatment of asthma (Labiris and Dolovich 2003). Asthma and chronic obstructive pulmonary disease (COPD) are two leading causes of death internationally within the umbrella of pulmonary disease, irrespective of gender. The death rate of respiratory diseases, including lung cancer, surpasses many other diseases affecting various parts of the body. According to the World Health Organization (WHO), the prevalence of lung diseases has steadily increased per annum. In the year 2000, the number of deaths caused by respiratory diseases was 23,849 (13,315 males and 10,533 females), by 2021, this number had substantially increased to 32,150 deaths (17,956 males and 14,191 females), with deaths being proportionally higher in males and females who fell in the age bracket of 55–75 onwards (WHO 2024).

Moreover, it is important to recognize that the pulmonary system is not only used for achieving localized effects where drugs can be deposited directly into the lung epithelium (requiring lower doses and resulting in rapid onsets of action) but has also been employed as a route for systemic drug delivery. The pulmonary route offers low enzymatic activity (both intracellular and extracellular within the system), which further enhances its potential high absorption rate and low degradation rate. In Fig. 1.1, it can be seen that the number of published papers (i.e. including research and review papers, book and monograph chapters, short communications, and patents) demonstrates an increasing trend of published articles in the area of pulmonary diseases. These data were collected using PubMed database, where the term "lung diseases" was used to search for trends. This figure not only displays the high number of published literature but also demonstrates interest in this topic. Hence, it is crucial to consider that pulmonary diseases are one of the six main categories of diseases worldwide, and only in England pulmonary diseases are the third most common cause of death, responsible for the deaths on an average of 68,000 people per year between 2013 and 2019, which is equivalent to one person every eight minutes (NHS 2024). Consequently, there has been an emergence of novel formulations at the micro and nanoscale, as well as an increasing number of traditional formulations and new drug delivery devices revolutionized for improved drug deposition and therapy. Generally, these inhalation devices are categorized into three main classes: dry powder inhalers (DPIs), pressurized meter dose inhalers (pMDIs), and medical nebulizers. These classifications of inhalation devices are based on the physical state of formulations (i.e., dry powder, solutions, or suspensions) that may be combined with each class of device to achieve the desired clinical effect. However, it is important to acknowledge that each category or class is further divided into different subcategories based on the device design, the composition of the formulation and its physical status (i.e. liquid or powder), and the

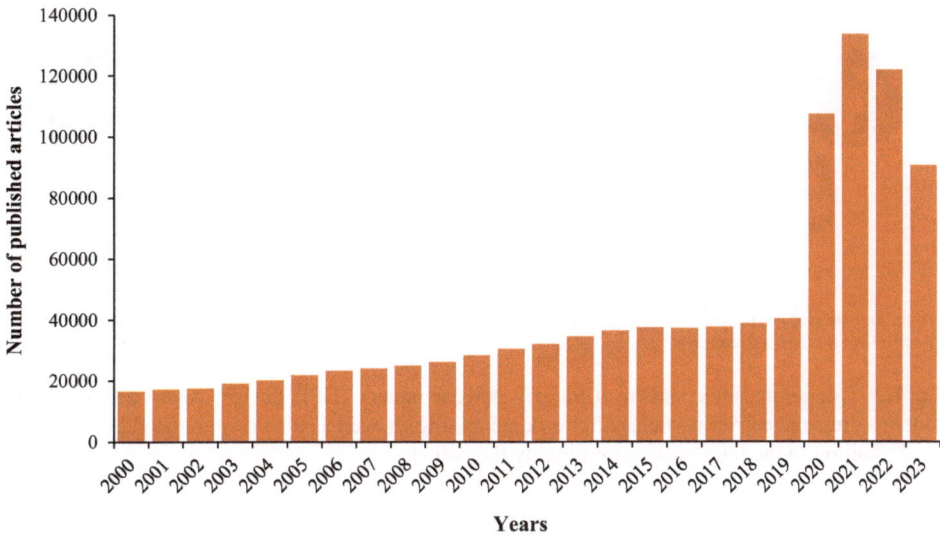

**Fig. 1.1** The number of articles published using the word "lung diseases" in their title in the PubMed database

portability of the device (Telko and Hickey 2005). Both DPIs and pMDIs contain formulations that release a specific quantity of formulation containing APIs (active pharmaceutical ingredients) upon each actuation, whereas in nebulizers, a solution or suspension of a drug can be loaded for aerosolization, and hence used in patients who require prolonged drug delivery or relatively large drug doses.

## 1.2   Types of Inhalation Devices

### 1.2.1   Dry Powder Inhalers (DPIs)

Dry powder inhalers (DPIs) deliver formulations in dry powder form to the pulmonary system in the absence of a propellant as an excipient. It is noteworthy that inspiratory flow rate is very important for successful deposition of an inhaled powder dose. DPI devices were launched in 1967, with the most basic prototypes described several decades ago (Goyal et al. 2017).

DPIs, where the main formulation type is comprised of micronized drug particles (i.e., less than 5 μm) and carbohydrate carriers e.g., lactose or glucose particles (i.e., more than 50 μm), combined via an ordered mix to avoid drug particles aggregation (Fig. 1.2). Large lactose particles are added to the powder formulations in order to enhance their flowability (Khan et al. 2016). Upon inhalation, the inspiratory flow rate may create turbulence and

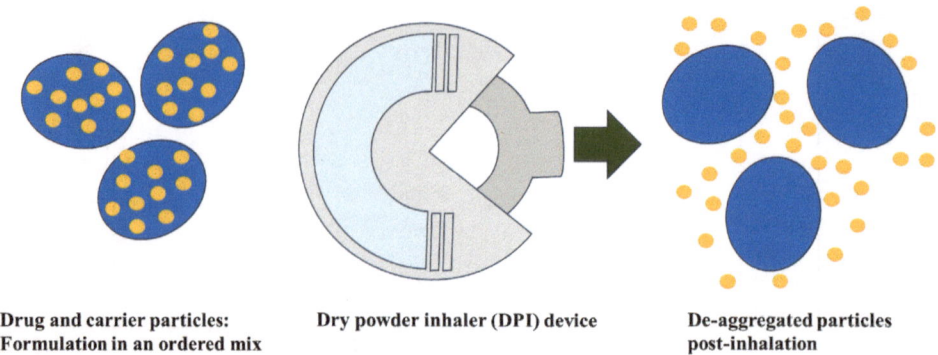

**Drug and carrier particles:**        **Dry powder inhaler (DPI) device**        **De-aggregated particles**
**Formulation in an ordered mix**                                          **post-inhalation**

**Fig. 1.2** Operational mechanism of the dry powder inhaler (DPI) device, where the formulation contains drug and carrier particles after ordered mixing, as well as the de-aggregated particles ready for deposition in the lung post-inspiration by the patient

shear force in the static powder formulation, leading to the liberation of micronized drug particles from the carrier particles, and hence deposition of the drug into the central and peripheral regions of the lungs. This type of drug deposition is directly related to particle size, where smaller and lighter (low aerodynamic) drug particles manoeuvre themselves in the pulmonary system and deposit via sedimentation (when the particle size range is 1–5 μm) and Brownian diffusion (i.e., particles are smaller than 1 μm). Large carrier particles (particles larger than 5 μm; commonly around 50 μm), due to their large size and density, would deposit in the upper respiratory tract in the oropharyngeal region via inertial impaction and be cleared by the mucociliary escalator system (Capanoglu et al. 2015; Telko and Hickey 2005).

In general, DPI devices are classified into three main categories. These are unit-dose inhalers, multi-dose inhalers, and multi-dose reservoirs. A number of various DPIs that are available on the market are highlighted in Table 1.1. In unit-dose inhalers, an individual capsule (made of gelatine or hydroxypropyl methylcellulose; generally, size 3) is filled with a powder formulation (Fig. 1.3). For example, the Rotahaler and Spinhaler were the only unit-dose inhalers available in the mid-1980s.

Multi-unit-dose inhalers contain blisters of many individual doses/capsules (Fig. 1.4). These include Diskhaler from GlaxoSmithKline (GSK), where upon actuation of a device, it leads to piercing the upper and lower surfaces of a blister via a needle, and upon inhalation, the blister is filled/dispersed with an air stream, efficiently dissociating the drug from the carrier, followed by its deposition in the lungs. After successful delivery, the disc automatically re-primes by rotating to expose the next dose in the blister pack.

The third category is multi-dose reservoirs, such as Turbuhaler, where the reservoir contains the total number of doses in bulk and dispenses a single dose through manipulation of the device prior to inhalation (Fig. 1.5). In this category of inhalers, upon rotating the conical cavity, the release of the drug into the airstream is triggered and dispersed into

**Table 1.1** Dry powder inhalers (DPIs) cover all three categories (i.e., unit-dose inhalers, multi-unit dose inhalers, and multi-dose reservoirs), with information related to each product available in the market

| DPIs device categories | Product name | Product manufacturer | Drug | Excipient(s) | Therapeutic condition | Mechanism | Shelf life (years) | Storage | Doses (capsules/ puffs) |
|---|---|---|---|---|---|---|---|---|---|
| Unit-dose inhalers | Foradil Aerolizer | Novartis Pharmaceuticals UK Ltd. | Formoterol fumarate | Lactose monohydrate | COPD and asthma | Selective $\beta_2$-adrenoceptor agonist | 10–15 days | ≤25 °C | – |
| | Arcapta Neohaler | Novartis Pharmaceuticals UK Ltd. | Indacaterol maleate | Lactose monohydrate | COPD and asthma | Long-acting $\beta_2$-agonist | 1 year | 15–30 °C | Box of 30 |
| | Seebri Breezhaler | Novartis Pharmaceuticals UK Ltd. | Glycopyrronium bromide | Lactose monohydrate, magnesium stearate | COPD and asthma | Bronchodilator, anticholinergics | 2 years | ≤25 °C | Box of 90 |
| | Spiriva | Boehringer Ingelheim Ltd. | Tiotropium bromide monohydrate | Lactose monohydrate | COPD and asthma | Anticholinergics | 2 years | ≤25 °C | Box of 30 |
| | Ultibro Breezhaler | Novartis Pharmaceuticals UK Ltd. | Glycopyrronium bromide, indacaterol maleate | Lactose monohydrate, magnesium stearate | COPD and asthma | Adrenergics in combination with anticholinergics | 2 years | ≤25 °C | Box of 96 |
| | TOBI Podhaler | Mylan | Tobramycin | 1,2-distearoyl-sn-glycero-3-phosphocholine, calcium chloride, sulfuric acid | Cystic fibrosis infection | Antibacterials | 4 years | * | Box of 56 |
| Multi-unit dose inhalers | Relenza Diskhaler | GlaxoSmithKline UK | Zanamivir | Lactose monohydrate | Influenza | Antiviral | – | ≤25 °C | Box of 20 |
| | Anoro Ellipta | GlaxoSmithKline UK | Vilanterol trifenatate, umeclidinium bromide | Lactose monohydrate, magnesium stearate | COPD and asthma | Adrenergics in combination with anticholinergics | 2 years | ≤30 °C | Box of 7 or 30 |

(continued)

**Table 1.1** (continued)

| DPIs device categories | Product name | Product manufacturer | Drug | Excipient(s) | Therapeutic condition | Mechanism | Shelf life (years) | Storage | Doses (capsules/puffs) |
|---|---|---|---|---|---|---|---|---|---|
| | Flovent Diskus | GlaxoSmithKline UK | Fluticasone propionate | Lactose monohydrate | COPD and asthma | Anti-inflammatory actions | 1 year | 20–25 °C | 1 foil of 60 blisters |
| | Incruse Ellipta | GlaxoSmithKline UK | Umeclidinium bromide | Lactose monohydrate, magnesium stearate | COPD and asthma | Anticholinergics | 2 years | ≤30 °C | Blister of 7 or 30 |
| Multi-dose reservoirs | Bricanyl Turbohaler | AstraZeneca UK Ltd. | Terbutaline sulfate | Lactose monohydrate | COPD and asthma | $\beta_2$-adrenergic agonist | 3 years | >30 °C | 120 |
| | ProAir Respiclick | Teva | Albuterol sulfate | Lactose monohydrate | COPD and asthma | $\beta$-adrenergic-receptor blocking agents | | 15–25 °C | 200 |
| | Oxis Turbohaler | AstraZeneca UK Ltd. | Formoterol fumarate | Lactose monohydrate | COPD and asthma | Selective $\beta_2$-adrenoceptor agonist | 2 years | * | 60 |
| | Pulvinal Salbutamol | Chiesi Ltd. | Salbutamol sulfate | Lactose monohydrate | COPD and asthma | Selective $\beta_2$-adrenergic receptor agonist | 1 year | ≤30 °C | 100 |
| | Easyhaler Budesonide | Orion Pharma (UK) Limited | Budesonide | Lactose monohydrate | COPD and asthma | Anti-inflammatory | 3 years | ≤30 °C | 200 |
| | Symbicort Turbohaler 100/6 | AstraZeneca UK Ltd. | Budesonide, formoterol fumarate dihydrate | Lactose monohydrate | COPD and asthma | Long-acting $\beta_2$-adrenoceptor agonist | 3 years | * | 60 or 120 |
| | Symbicort Turbohaler 100/6 | AstraZeneca UK Ltd. | Budesonide, formoterol fumarate dihydrate | Lactose monohydrate | COPD and asthma | Anti-inflammatory, selective $\beta_2$-adrenoceptor agonist | 3 years | * | 60 or 120 |

* This medicinal product does not require any special storage conditions. Keep the container/cap tightly closed

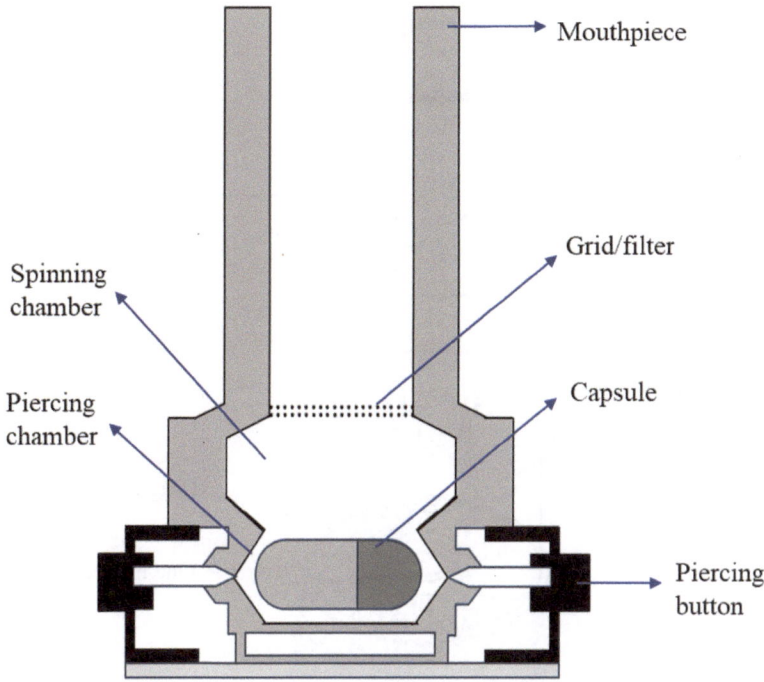

**Fig. 1.3**  A diagram of a simple single capsule-based unit-dose inhaler as an example of a DPI

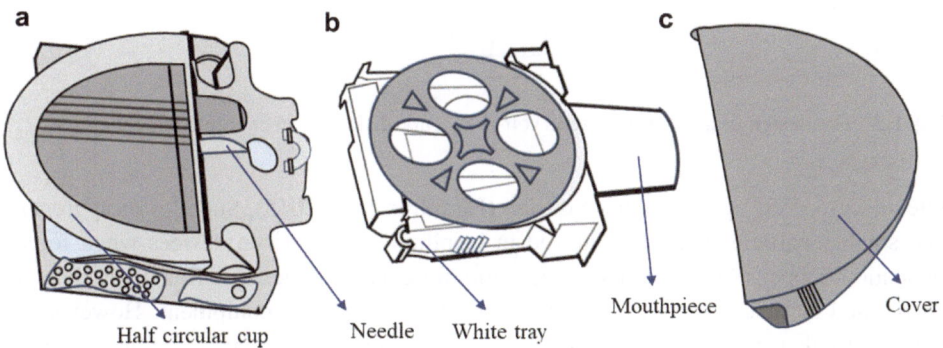

**Fig. 1.4**  A multi-unit dose inhaler (i.e., Diskhaler) is an example of DPIs, illustrating **a** a main body, **b** a sliding tray, and **c** a cover slip

channels for inhalation, which are helical in shape and induce turbulent airflow. Some examples of multi-dose reservoirs are the Easyhaler, Clickhaler, and Pulvinal.

Performance of DPI formulations depends primarily upon particle shape, size, porosity, aerodynamic diameter, powder flow properties, drug-carrier interaction, inspiratory flow

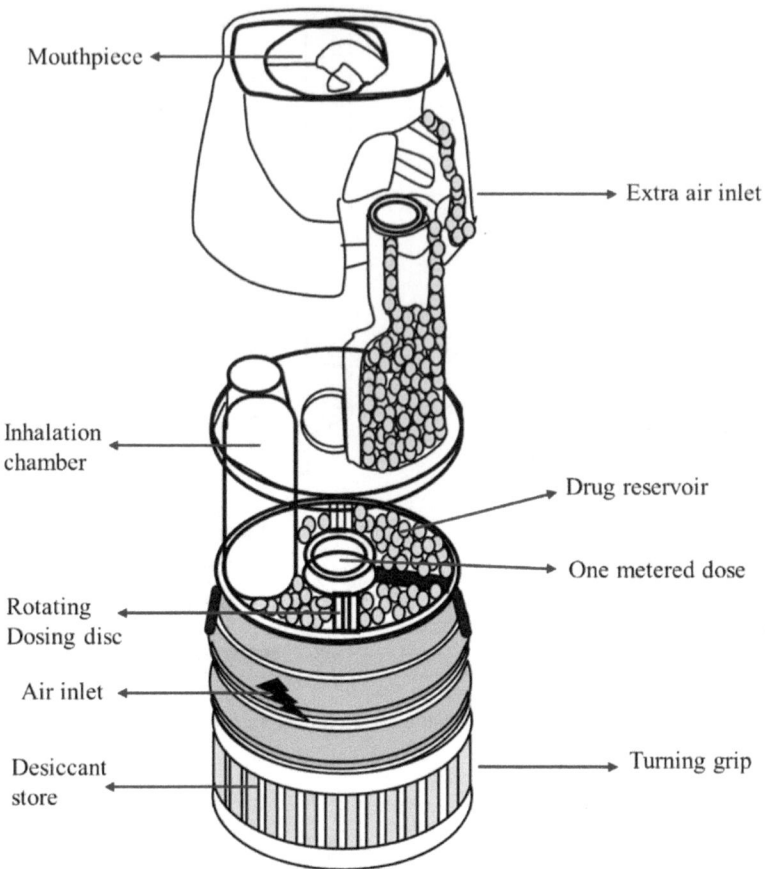

Mouthpiece

Extra air inlet

Inhalation
chamber

Drug reservoir

One metered dose

Rotating
Dosing disc

Air inlet

Desiccant
store

Turning grip

**Fig. 1.5** The design of a multi-dose reservoir (i.e., Turbuhaler) as an example of DPIs

rate, disease state, age, and device design (Hamilton et al. 2015). Superior disaggregation and smaller particles may be achieved in inspiratory flow-driven devices when forceful inhalation happens by the patient. The optimum characteristics in terms of particle size, shape, and surface properties, can be attained using milling equipment. However, it is noteworthy that micronized drug particles tend to adhere to surfaces and are difficult to handle due to electrostatic charges and Van der Waals forces. Therefore, the incorporation of large carbohydrate carrier particles solves this problem, since smaller drug particles would adhere to the surface of the large carrier particle during mixing. Importantly, this interaction or adhesion should not be overly powerful; otherwise, this may retard the detachment of drug particles from carriers during inhalation (Ezzati Nazhad Dolatabadi et al. 2015). The dispersion of powder mix can be attained using different mechanisms by various delivery devices, including the Aspiror, Enhance (via a vacuum chamber or pressurized air), Airmax (via a cyclone chamber or circulation), Easyhaler (via a venturi nozzle), and Turbuhaler (via channels with spiral inserts or release channels).

Furthermore, there are various factors that can affect the deposition or dose delivery by patients using DPIs. Some of these factors include drug-carrier interactions, drug and carrier particle size and shape, flowability of formulation ingredients, aerosol generation and delivery from the inhaler device, and patient inspiratory flow rate and inhalation technique. For effective drug delivery, where the maximum amount of dose should reach the targeted areas of the pulmonary system, a proper inhalation technique is required. Based on age or poor inhalation flow rate, improvements in efficacy can be made by employing add-on devices such as spacers. However, a number of pharmaceutical factors are also important for the development and advancement of new inhalation device designs. These include the approval in accordance with pharmacopoeia and regulatory requirements. Formulation and dose delivery fractions that should be $<5$ $\mu$m are also considered. Along with pharmaceutical factors, clinical applications are equally significant in order to achieve the required therapeutic effects; the devices must be easy to use, target specific areas of the lungs, training should be provided in their use, and they a minimum inspiratory flow rate is applied by the patient. In pharmaceutical formulations, advanced drug delivery systems have been employed in recent years to improve delivery effectiveness by encapsulating drug molecules within micro or nanoparticles (Khan et al. 2023).

## 1.2.2  Pressurized Metered Dose Inhalers (pMDIs)

Pressurized metered dose inhalers (pMDIs) are hand-held portable devices that are also commonly referred to as puffers or metered dose inhalers, and are used for pulmonary illnesses such as chronic obstructive pulmonary disease (COPD) and asthma (Dolovich and Dhand 2011). Patients who suffer from such pulmonary diseases, previously used to take medication in a glass bulb automizer, which is often associated with leakage of the medication. A similar story is also related to a 13-year-old asthmatic girl named Suzi (the daughter of Dr George Maison, president of Ricker Laboratories), who would administer her formulation via a device akin to a hairspray can (Rubin and Fink 2005). Dr George Maison, along with his colleagues, started working on the project and introduced the first pMDI in 1956. This device contained a bronchodilator in combination with a propellant and was developed with an old freezer, a case of soda bottles, and a bottle capper (Fink 2000). This pMDI was significantly more advanced than the previous devices, and since then, in the coming years, its popularity has substantially increased.

Since the pMDI was conceptualised, within a few months, the first pMDIs were developed using salts of isoproterenol and epinephrine within a 10 ml amber vial and a metering valve of 50 $\mu$l for perfume aerosol (which also contained a moulded nozzle in a 3-inch-long plastic mouthpiece) (Fink 2000). The effectiveness of these first pMDIs showed promise when they were clinically trialled at the Long Beach Veterans Administration

Hospital. This was then followed by the approval of a new drug in the pMDIs. In 1957, the original bronchodilator used was substituted with a dispersion of micronized drug in a propellant containing surfactant exhibiting further promising results (Rubin and Fink 2005). However, many former pMDI users complained that they did not get as much medication. This misunderstanding was related to the taste of previous formulations that contained alcohol, and with patients conflating this with amount of active drug within the formulation. Since the introduction of the first pMDI, for the last 68 years, it has been used as a backbone for pulmonary disorders, subsequently many changes have been made in its design and formulations to facilitate patient use and therapeutic efficacy (Table 1.2).

The composition of a pMDI is mainly made up of a canister (which contains the formulation), a metering valve (which releases a precise dose of formulation each time), and an actuator (the main body containing the metering valve and holding canister) (Fig. 1.6). Each component of the pMDI is extremely important and may affect the process of aerosol formation and subsequent drug delivery to the pulmonary system. The pMDI canister is largely comprised of aluminium with or without an inner coating (earlier canisters were made of glass vials coated with plastic) and is mounted in an actuator (Heyder et al. 2017). Aluminium canisters are lighter than stainless steel and are considered superior to glass vials, as they are less fragile, more compact, and light-proof. The formulation (either solution or suspension) contains an active pharmaceutical ingredient (API) and a small amount of one or more excipients and a co-solvent, but mostly contains a large amount of one or more propellants. The formulation is stored under pressure within the canister; this pressure remains almost constant as long as the liquid is present within the reservoir. pMDIs canisters have thicker walls based on the pressure inside and hold a standard volume of formulation ranging from 10 to 22 ml. The inner wall of the canister is coated in order to prevent drug degradation and interaction, as well as decrease particle deposition and adhesion on the canister wall. Various coating techniques and coating materials are utilised, but selection is dependent upon the type of drug formulation in order to overcome inconsistent dosing and stability issues. The most commonly used coating materials are based on fluorocarbon polymers, which contain a lower ratio of carbon and a high ratio of fluorine (for example, polytetrafluoroethylene and perfluoroalkoxyalkylene) and are chemically inert. A standard metal plate coating is used prior to manipulating the material into the components/spraying, and then electrostatic dry-powder coating followed by drying (where a high temperature of up to 400 °C is used). An alternative method that may be employed for coating is referred to as glass plasma. This technique does not require high temperatures (i.e., less than 45 °C for polymeric surfaces and less than 75 °C for metal surfaces) but needs to be conducted under vacuum. This method of coating provides a very thin protective layer and reduces deposition, degradation, and corrosion.

The stability of colloidal particles in pMDI formulations was characterized using a surface component approach based on two types of interaction, including a polar Lewis acid–base component and a non-polar Lifshitz–van der Waals component. A great difference in adhesion was found between different canister materials and different drugs,

**Table 1.2**   Pressurized meter dose inhalers (pMDIs) in solution or suspension form contain information related to each product available in the market

| Product name | Device and formulation type | Product manufacturer | Drug | Excipient(s) | Therapeutic condition | Therapeutic class | Shelf life | Storage | Doses (actuations) |
|---|---|---|---|---|---|---|---|---|---|
| Flutiform | Valve, canister, actuator with dose counter (suspension) | Napp Pharmaceuticals Ltd. | Fluticasone propionate, formoterol fumarate dihydrate | Sodium cromoglicate, ethanol anhydrous, apaflurane HFA 227 | Asthma | Corticosteroid and a long-acting β₂-agonist | 2 years | ≤25 °C | 20 |
| Symbicort Turbohaler | Valve, canister, actuator with dose counter (suspension) | AstraZeneca UK Ltd. | Budesonide, formoterol fumarate dihydrate | Apaflurane (HFA 227), povidone, macrogol | Asthma | Corticosteroid and long-acting β₂-adrenoceptor agonist | 2 years | Room temperature | 60 or 120 |
| Qvar 100 | Valve, canister, actuator (solution) | Teva UK Ltd. | Beclometasone dipropionate | HFA-134a (norflurane), ethanol | Asthma | Glucocorticoid | 3 years | ≤25 °C | 100 or 200 |
| Clenil Modulite | Valve, canister, actuator (solution) | Chiesi Ltd. | Beclometasone dipropionate | HFA-134a, ethanol, glycerol | Asthma | Glucocorticoid | 3 years | ≤30 °C | 200 |
| Atrovent | Valve, canister, actuator with dose counter (solution) | Boehringer Ingelheim Ltd. | Ipratropium bromide monohydrate | 1,1,1,2—tetrafluoroethane (HFA-134a), ethanol anhydrous, purified water, citric acid anhydrous | COPD and asthma | Anticholinergics | 3 years | ≤25 °C | 200 |

(continued)

**Table 1.2** (continued)

| Product name | Device and formulation type | Product manufacturer | Drug | Excipient(s) | Therapeutic condition | Therapeutic class | Shelf life | Storage | Doses (actuations) |
|---|---|---|---|---|---|---|---|---|---|
| Ventolin | Valve, canister, actuator with dose counter (suspension) | GlaxoSmithKline UK | Salbutamol sulfate | HFA 134a | Asthma | Selective $\beta_2$-andrenoreceptor agonists | 2 years | <30 °C | 200 |
| Alvesco | Valve, canister, actuator with dose counter (solution) | Zentiva | Ciclesonide | Norflurane (HFA-134a), ethanol | Asthma | | 3 years | * | 60 or 120 |
| Atimos Modulite | Valve, canister, actuator (solution) | Chiesi Ltd. | Formoterol fumarate dihydrate | Norflurane (HFA-134a), ethanol anhydrous, hydrochloric acid | COPD and asthma | Selective $\beta_2$-adrenoreceptor agonists | 2.5 years | 2–8 °C | 50 or 120 |

* This medicinal product does not require any special storage conditions

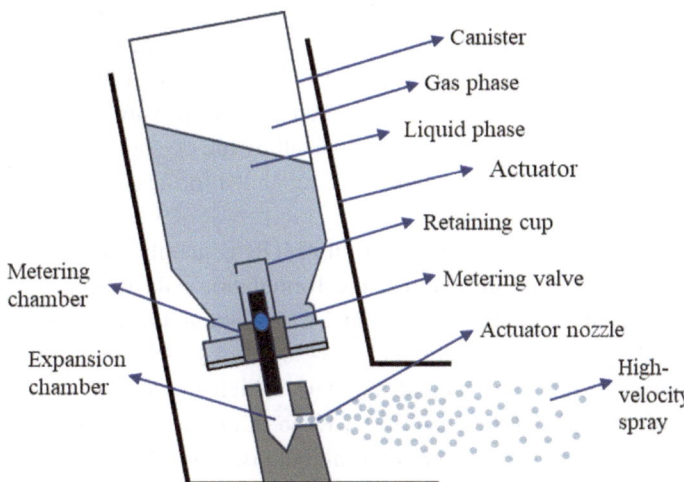

**Fig. 1.6** A schematic representation of pMDI illustrating a canister enclosing the drug dissolved or dispersed in a pressurized propellant. Upon actuation, a precise volume is released from the nozzle, resulting in immediate evaporation of the propellant to leave a precise dose of the drug for inhalation

and the wall of an aluminium canister coated with a polymer coating greatly reduced the force of interaction. A water-free propellant was employed, as water is a polar solvent and can significantly influence the polarity of the propellant and hence particle adhesion. It was also advocated that canister surfaces (i.e., surface morphology) play a pivotal role in particle adhesion (Traini et al. 2005, 2006; Young et al. 2003).

In addition to the importance of design and coating materials, manufacturing methods are equally important, especially canister filling. Cold and pressure filling are the two main methods utilised (Haswell 2020; Vallorz et al. 2019; Errington 2012). These methods are different from each other and require different formulations. In the cold filling method, a drug suspension or solution is prepared at room temperature using a suitable solvent. Simultaneously, a pre-cooled vessel is used to convert the propellant into a liquid. After mixing both, the cooled mixture is then transferred into an open canister and closed after filling with a head (containing a valve system to release a single dose). This is followed by crimping the head with a rubber seal and measuring the weight (to ensure the correct and consistent quantity of formulation). The final task is to perform the leak test of the filled canister with water. Pressure filling is further divided into two types: a one-step process and a two-step process. In the one-step process, both propellant and drug formulations are mixed together under pressure. This mixture is placed in the canister by forcing it through the valve system (already crimped to the canister) in the reverse direction. In the two-step process, firstly, the drug suspension or solution is placed in the canister (prior to crimping the head), and secondly, the propellant, via the reverse direction, is forced through the valve system into the canister. Both methods have their own advantages when it comes to formulation design and type (Errington 2012). Cold filling helps in checking the optimization and homogeneity of final formulations easily after mixing the propellant with the concentrated drug suspension before filling into the canister. Moreover, this method also aids in controlling the crystallization and particle size of the drug used. Pressure-filling is highly recommended for large-scale manufacturing of pMDI formulations.

The formulation (solution or suspension) in pMDIs is mainly comprised of surfactants (to prevent cake formation and stabilize the formulation), flavouring agents, preservatives, a drug content of approximately 1%, and a high amount of propellants (circa 80%). Furthermore, pMDIs contain approximately 15–30 ml of formulation, are designed to be robust and light in weight, and should maintain a high interior pressure of 3–5 atmospheres (i.e., 300–500 kPa) (Newman 2005; Khan et al. 2016). As propellants are used at a relatively high percentage in pMDI formulations, chlorofluorocarbons (CFCs, with the tradename of Freon) are the most commonly employed propellants. The propellant must be sealed tightly in order to maintain the high pressure and to retain the propellant in liquid form so that the drug remains dispersed or solubilized (Hugh 2003). In the preceding decades all pMDIs incorporated CFCs as propellants. Their use as propellant was reported to cause ozone layer depletion, and hence, this was the reason for their consequent ban in the mid-1980s (where CFC was used in common aerosol products) in the Montreal Protocol established in 1987. CFC was not banned until its use was ceased in 1996. The main reason was the lack of introduction of new alternative propellants as well as the high cost of their toxicity tests. Therefore, alternative propellants, such as hydrofluoroalkanes (i.e., HFA-134a) were introduced. HFA propellants were found to be non-toxic, non-flammable, and chemically inert. With the change in the propellant, the metering valve was also changed to have a much smaller aperture to produce much smaller and finer particles. HFA directly affects the particle size of some drugs, such as beclometasone dipropionate, which is soluble in HFA but remains a suspension in CFC. The velocity of HFA-propelled aerosols is lower when CFC is used; HFA also produces smaller particle sizes and gentler plume formation, which is attributed to lower drug deposition in the oropharyngeal region; hence, HFA is identified as an efficient and reliable propellant in comparison to CFC.

Upon actuation, formulations are released through a metering valve for inhalation. The actuator is the main body and comprises a metering valve (to release a consistent dose of formulation), a mouthpiece (to allow the patient to inhale the formulation), and a nozzle (to control the particle size distribution). The actuator also holds the canister, which works as a protective cover (Smyth et al. 2006). Any small changes in the canister may lead to an alteration in the output and aerosol characteristics. The efficiency and performance of pMDIs are mainly dependent on the metering valve. The output volume, or aerosolized formulation, ranges from 25 to 100 $\mu$l and releases approximately 20 $\mu$g–5 mg of drug per actuation. In adults, the deposition of formulations is anticipated to be 10–25%, with high variability from patient to patient based on their inhalation technique. The generation of aerosol from the pMDI takes about 20–30 ms, where the aerosolized droplets released from the canister start upon the vaporization of the propellant and hence leave the actuator through the nozzle in the form of a plume (and remain as the propellant evaporates). The particles produced from the vaporization of the propellant are initially 35 $\mu$m but this size rapidly decreases (attributed to evaporation as the plume travels away from the nozzle), and the mass median aerodynamic diameter (MMAD) of the resultant particles from the pMDIs becomes 2–6 $\mu$m, with deposition in the lungs ranging from 10 to 20% ($\sim$ 80% deposit in the oral cavity). However, for maximizing formulation deposition in the "deep lung", patient inhalation technique and patient education regarding proper use of the device are of paramount importance. In pMDIs, coordination between actuation and inhalation take place simultaneously. To overcome the lack of coordination issue, spacers have been introduced. There are spacers that are specially designed for children and have a volume of 200–300 ml to aid formulation inhalation. In these instances, spacers are usually connected with pMDIs, which provide additional volume to capture aerosols, also aiding in propellant evaporation, and reducing formulation velocity and droplet size, hence improving the inhalation of aerosol droplets. With further advancements in pMDI devices, dose counters were introduced (i.e., containing 100–200 doses) as additional devices to count the number of doses and inform the patient of the number of remaining doses.

### 1.2.3  Medical Nebulizers

In response to the identified pitfalls of both DPIs and pMDIs, medical nebulizers were introduced. Nebulizers produce continuous aerosol formation, making it convenient for patients to inhale drug formulations via aerosol droplets. With the use of these delivery devices, potential growth was found. The number of articles using nebulizers is highlighted in Fig. 1.7, where the number of articles published increased from year to year.

In the year 1912, the first inhaler device in the form of a glass nebulizer was introduced for the treatment of asthma (when Ephraïm employed adrenaline for asthma treatment) (Price et al. 2011; Nicolini et al. 2010). In the case of respiratory diseases, nebulizers are considered valuable and efficient tools, possessing the ability to convert suspensions or solution formulations (containing antibiotics, anticholinergics, mucolytic agents, bronchodilators, and corticosteroids) into small droplets (with a wide polydispersity index). The delivery of aqueous droplets into the pulmonary system has been established for decades (Dhand 2003). Additionally, nebulizers can deliver drugs in high doses in a continuous manner when compared to their counterparts, DPIs and pMDIs, which require multiple doses. For example, individuals suffering from chronic obstructive pulmonary disease (COPD) or asthma take high doses of terbutaline relievers, ipratropium, and salbutamol, which are comparable to 4–6 puffs of inhaler devices.

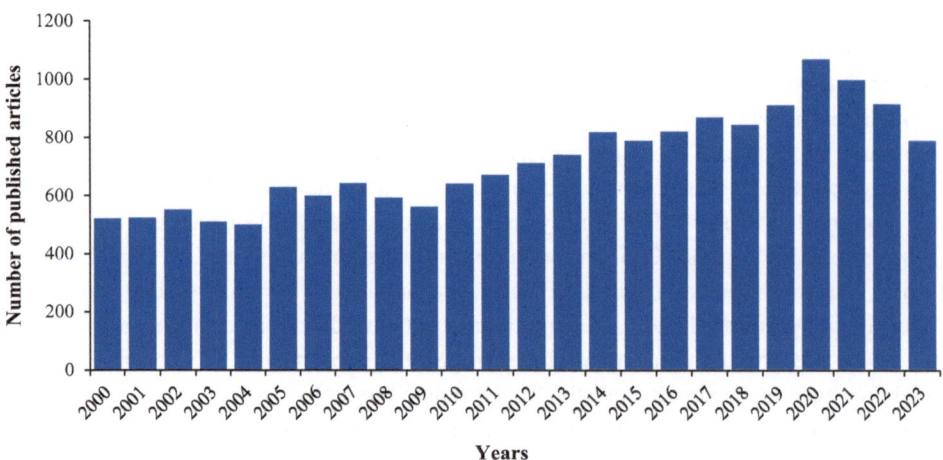

**Fig. 1.7**  The number of articles published using the words "nebulizer, nebuliser" in their title in the PubMed database

Nebulizers are considered the oldest form of aerosol generation device and can be employed for any age and many diseases. It is noteworthy that drug delivery via nebulization is more convenient for elderly patients and children, attributed to the minimum requirements in terms of coordination and effort for inhalation. The efficiency of aerosol generation via nebulization is directly related to nebulizer design, patient breathing pattern, and formulation physicochemical characteristics such as viscosity and surface tension. Nebulizer design plays a crucial role since it significantly influences the size of aerosol droplets and aerosol output. Moreover, droplet size is not dependent on the fill volume in the reservoir; the fill volume may only increase the aerosolization of the drug proportion for inhalation. A change in temperature during nebulization may affect the formulation output, notably causing crystallization of hydrophobic drugs; hence, a change in the physicochemical properties may also affect the size of the droplets from the nebulized fluid.

The literature documents the employment of nebulizers for a number of micro and nano lipidic and polymeric nanoparticle formulations, including; niosomes and proniosomes (Elhissi et al. 2013; Benedini and Messina 2022), liposomes and proliposomes (Khan et al. 2018; Elhissi 2017), emulsions (Hecker et al. 2015; Orizondo et al. 2016), solid lipid nanoparticles (Leong and Ge 2022), nanostructured lipid carriers (Pindiprolu et al. 2020; Almurshedi et al. 2021), and polymeric nanoparticles (d'Angelo et al. 2015). Smaller particles (i.e., 1–5 μm) possess a higher tendency to reach the central and alveolar regions in the lungs. Drug suspensions, where drug particles are suspended in an aqueous solvent, may also facilitate the delivery of suspensions to the pulmonary system via aerosol droplets. Formulations with relatively high viscosities may take longer to nebulize. Similarly, upon reducing the surface tension, nebulized formulations have a tendency to produce aerosols of smaller size (McCallion et al. 1996).

In general, nebulizers are categorized into three main types, these are air-jet nebulizers, ultrasonic nebulizers, and vibrating mesh nebulizers. Each of the types is then further divided into sub-categories.

### 1.2.3.1 Air-Jet Nebulizers

Air-jet nebulizers are also referred to as pneumatic or compressor nebulizers (Laube and Dolovich 2014). These nebulizers are typically comprised of a mouthpiece, a plastic bottle (which also contains a reservoir to hold the formulation), a nozzle (also referred to as a venturi, typically 0.3–0.7 mm in diameter), a baffle system, and a compressor (Fig. 1.8). The mouthpiece is usually connected to a mask to improve the inhalation of aerosol droplets. In general, after pouring the aqueous formulation (i.e., solution or suspension form of a few millilitres, i.e., 2–5 ml) into the nebulizer reservoir, compressed air is applied via a compressor (with the aid of a plastic tube attached at the bottom of the jet-nebulizer) to convert the liquid medication into inhalable fine droplets (Kendrick et al. 1997). When using compressed air with high velocity, this enters the plastic tube from the bottom orifice of the tube and exits through the small orifice or nozzle. Upon passing air with high velocity through the nozzle, the "venturi" may cause the formation of negative pressure (an area of low pressure at the outlet of the adjacent feed tube) and draw up the liquid formulation from its reservoir via a feed tube (referred to as the "Bernoulli" effect). The withdrawn liquid forms a fine filament in the air stream and collapses into droplets (with a wide size distribution) because of surface tension. Typically, the generated droplets are in the size range of 15–50 μm, which is referred to as the "primary aerosol".

It is important to note that the baffle is located above the venturi nozzle, which principally works as a "sieve", allowing only smaller droplets (the secondary aerosol) to pass through for inhalation (Labiris and Dolovich 2003; McCallion et al. 1996). The non-respirable, larger droplets hit the baffle or internal walls, in order to become smaller and capable of escaping the baffle as "secondary" aerosol. Also, the larger droplets after collision may deflect back into the nebulizer reservoir for re-aerosolization. This is associated with a decrease in formulation temperature (by ~ 10–15 °C after several minutes) because of solvent evaporation driven by the compressed air. After completed nebulization in 5–20 min (when there is no further aerosol generation), a proportion of the concentrated formulation remains in the nebulizer reservoir (also termed "dead volume" or "residual volume"). Upon using 3–5 ml of formulation for nebulization, as opposed to 2 ml, the aerosolized drug formulation is much higher for inhalation (and will result in a longer nebulization time), and the formulation that stays in the nebulizer reservoir is around 0.5–1.5 ml (Khan et al. 2020a, 2021a). The higher the dead volume, the lower the drug availability for patients.

There are two common designs of the venturi nozzle in the air-jet nebulizers (Fig. 1.9) (Hess 2000). The first type is the external mixing design, where both medication fluid and

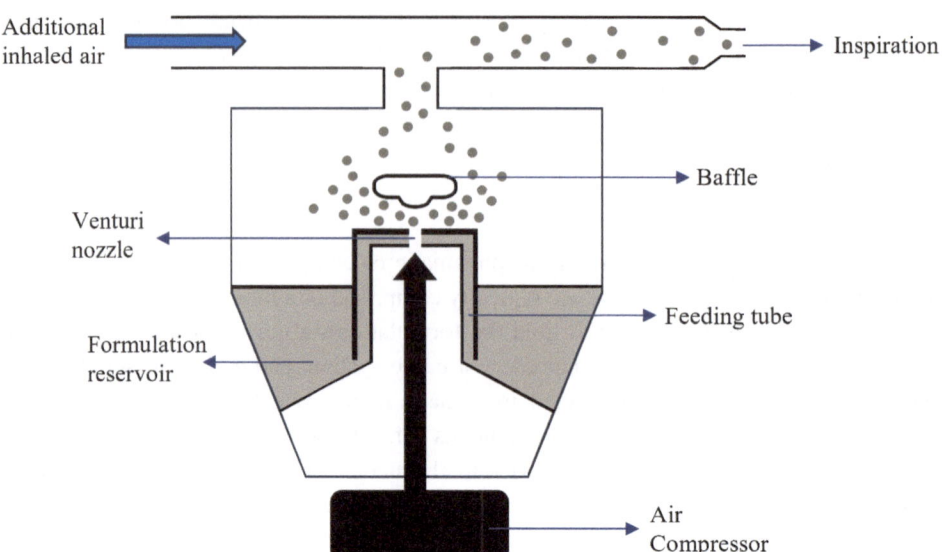

**Fig. 1.8**  A conventional air-jet nebulizer containing a mouthpiece, a plastic bottle, a nozzle, and a baffle with an electrically operated compressor

compressed gas pass separately (i.e., without mixing), and upon exit, they mix simultaneously. The second type is an internal mixing design, where both compressed gas and medication fluid mix and interact before exiting the venturi nozzle.

Jet-nebulizers are easy to use and very economical; however, during the inhalation by the patient, some drug loss may occur through exhalation. Moreover, after complete nebulization, a high residual volume stays in the plastic bottle, which can be reduced by reducing the concentration of solute in the formulation (Hess et al. 1996). Jet nebulizers have been in use for long decades, but with the incorporation of new designs and technologies, the efficiency of jet nebulizers has improved significantly, including breath-enhanced jet-nebulizers and breath-actuated jet-nebulizers.

### Breath-Enhanced Air-Jet Nebulizers

The operation mechanism of this type of nebulizer is the same as that of conventional air-jet nebulizer, and it generates continuous aerosol during its operation. These nebulizers have demonstrated higher output during the inhalation process. This means that drug loss during inhalation and exhalation by patients is decreased when using breath-enhanced air-jet nebulizers as compared to conventional air-jet devices. Some examples of breath-enhanced jet nebulizers are the LC Plus, LC Sprint, and LC Star nebulizers (Pari, Germany). This change in the output efficiency is directly related to the design of the breath-enhanced jet-nebulizers (Fig. 1.10). Breath-enhanced jet nebulizers contain an extra valve located near the mouthpiece. The mouthpiece can be connected to the mask,

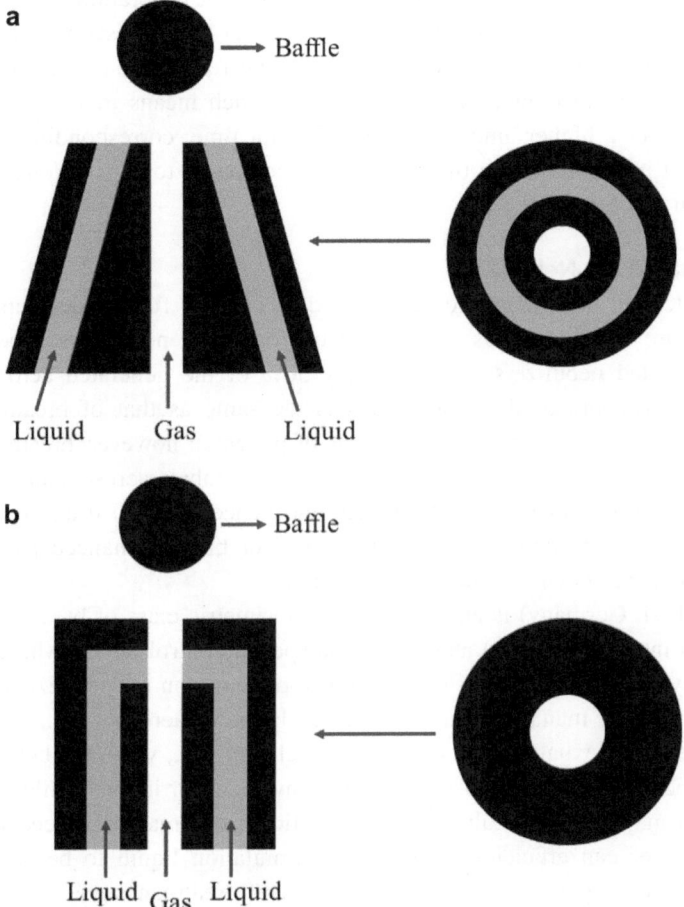

**Fig. 1.9** The designs of the venturi nozzle in air-jet nebulizers during nebulization, illustrating the mixing process of compressed air/gas with formulation liquid using: **a** external mixing, and **b** internal mixing with the front view of the nebulizer tip geometry

from which the patient can inhale the aerosol. During the operation and patient inhalation and exhalation, this valve stays closed during inhalation and opens during exhalation. Additionally, there is an open vent on the top side of the nebulizer for air, which opens during inhalation and closes during exhalation. Therefore, the additional auxiliary air flow via the open vent moves the generated aerosol cloud out of the nebulizer with each inspiration, which leads to shorter nebulization times (i.e., increasing convenience for patients). Both the open vent and valve work together, but oppositely, in order to minimize drug/aerosol losses. These additional parameters in the design of the jet nebulizer may enhance the air flow rate and improve small droplet generation, hence their deposition in the lower respiratory tract. It is noteworthy that highly-viscous drug formulations require high-power compressors, whereas for normal formulations, an economical compressor is sufficient to reduce nebulization time, which means that breath-enhanced jet nebulizers deliver a higher amount of drug per unit time, corresponding to higher total drug delivered in the minimum time span when compared to conventional jet nebulizers (O'Callaghan and Barry 1997).

**Breath-Actuated Jet Nebulizers**

Breath-actuated jet nebulizers are advanced devices with further development in their designs and are also known as intermittent or dosimetric nebulizers. Upon using continuously operated nebulizers, approximately 50% of the generated aerosol is wasted. The working principle of these nebulizers is the same as that of breath-enhanced jet nebulizers (inhalation and exhalation patterns of patients); however, breath-actuated nebulizers deliver aerosolized droplets only when patients inhale aerosols and not when they exhale. This may eliminate aerosol wastage and hence improve drug delivery approximately threefold when compared to conventional or breath-enhanced jet nebulizers, in addition to reducing environmental contamination.

Pari LL (Pari, Germany) is an example of a dosimetric class of breath-actuated nebulizer (i.e., an intermittent type) and contains a special control button where coordination is required by the patient in order to start and stop the button for nebulization (Fig. 1.11). This button must be managed manually for inhalation of aerosols (i.e., when the button is switched on) and exhalation to stop aerosol delivery (i.e., when the button is switched off). This type of nebulizer is very useful and convenient for infants, children, and elderly patients who may have difficulty using conventional or breath-enhanced jet nebulizers. Such a nebulizer can efficiently control the formulation liquid to be aerosolized (i.e., reduce aerosol wastage) and hence reduce the overall treatment time.

AeroEclipse (Trudell Medical International, London, Ontario, Canada) is another example of a breath-actuated jet nebulizer (Fig. 1.12). This nebulizer relies upon a new spring-loaded technology with a one-way valve design (i.e., a sensor technology). A green button is located at the top of the nebulizer, which moves up and down during patient inhalation and exhalation of aerosols. Upon inhalation by a patient, the nebulizer baffle or actuator piston (located directly above the orifice or nozzle) moves down, resulting in aerosol generation, whereas upon exhalation, the nebulizer baffle moves back to the original resting position and completely ceases aerosol production until the next breath or inhalation by the patient. Therefore, during exhalation, there is no drug waste, resulting in a longer nebulization time when compared to the counterpart nebulizers.

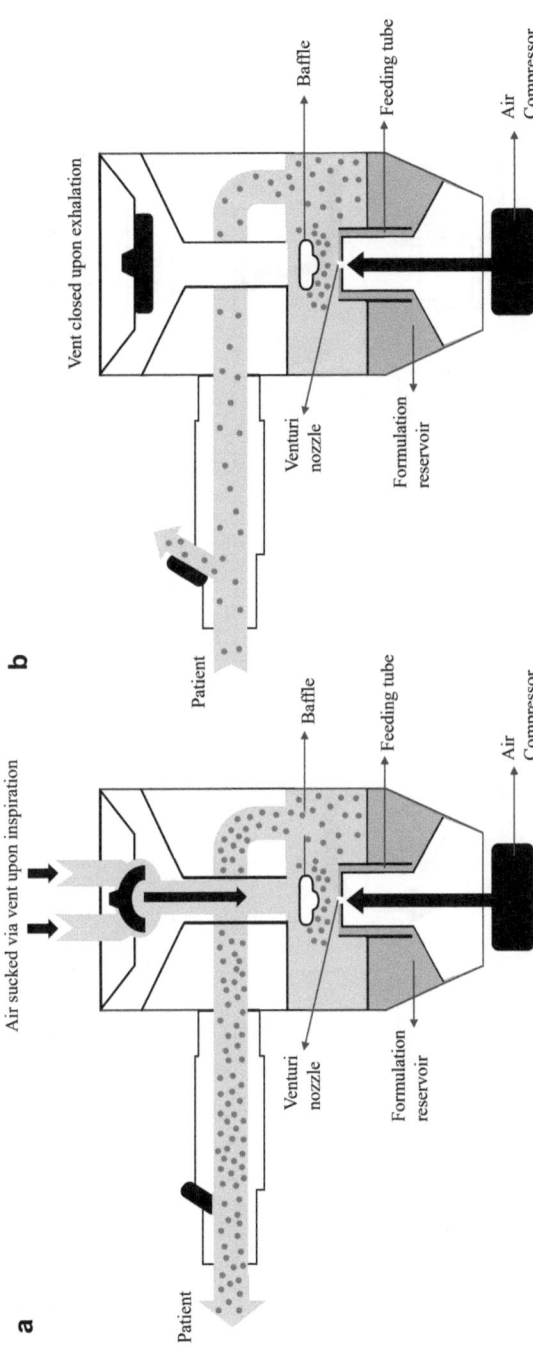

**Fig. 1.10** A schematic representation demonstrating the principle of operation of the Pari LC Plus nebulizer (an example of a breath-enhanced jet nebulizer). **a** During inspiration, the valve present at the top of the nebulizer opens, allowing the patient to inhale more aerosol droplets of the formulation along with extra air. **b** Upon expiration, the valve at the top of the nebulizer closes, allowing the aerosol droplets to exit from the one-way valve near the mouthpiece

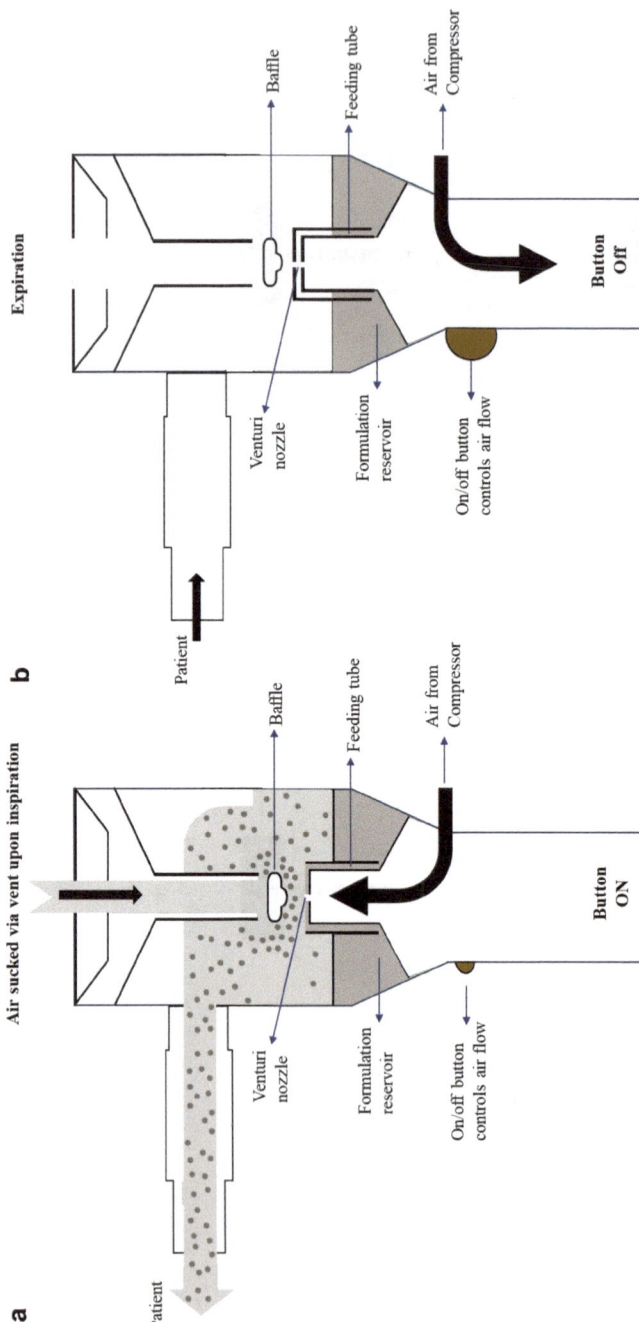

**Fig. 1.11** A graphical demonstration and principle of operation of the Pari LL nebulizer (an example of a dosimetric type of breath-actuated jet nebulizer). **a** Upon switching on the button, aerosols are generated as compressed air passes through the nebulizer. **b** Aerosol generation stops when switching off the nebulizer button

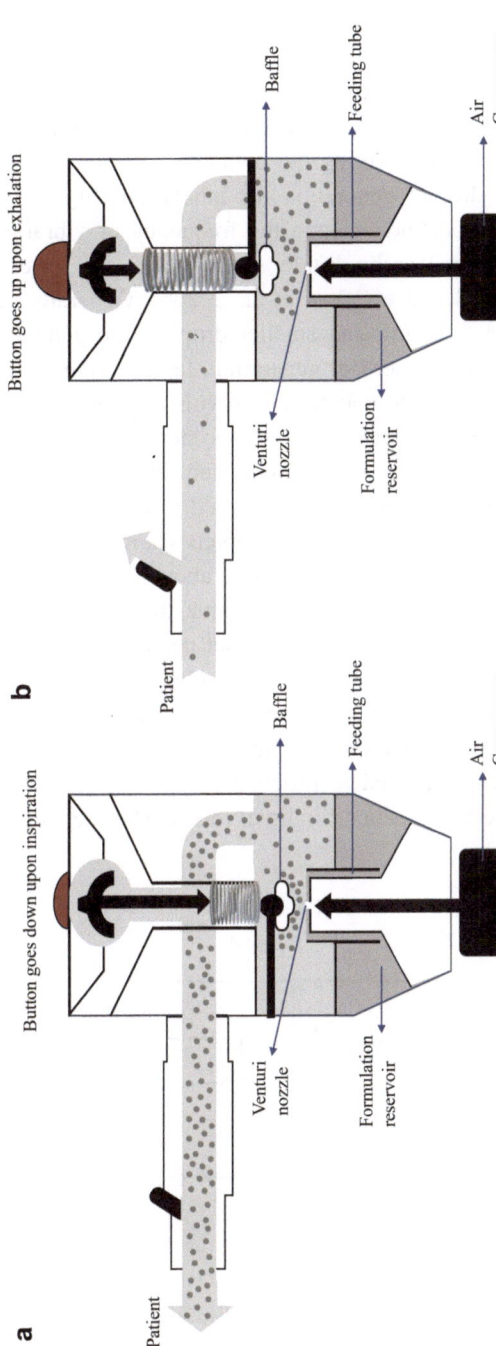

**Fig. 1.12** The operation mechanism of the AeroEclipse nebulizer (an example of a breath-actuated jet nebulizer). **a** Upon inspiration, the actuator piston/spring (the nebulizer baffle) and green button move down and release aerosol droplets for inhalation. **b** During expiration, the piston and green button move up and cease aerosol generation

### 1.2.3.2   Ultrasonic Nebulizers

Ultrasonic nebulizers possess many properties that are similar to those of jet nebulizers; however, the mechanism of generation of aerosols is distinctly different. Ultrasonic nebulizers generate ultrasound waves via piezoelectric crystals located below the cup that holds the nebulization fluid. These crystals produce high frequency vibrations within the cup containing the formulation, transforming the contained aqueous formulation into a fountain, resulting in aerosol droplet generation for inhalation (Fig. 1.13). Smaller or larger droplet size formation is dependent upon the frequency of vibration. The higher the frequency of vibration, the smaller the droplet size, whereas the lower the frequency of vibration, the larger the droplet size (Khan et al. 2020b). Generally, larger droplets are formed at the apex of the fountain, and smaller droplets are generated at the bottom (Elhissi and Taylor 2005). Upon aerosol generation, large droplets (i.e., the primary aerosol) may be deflected back into the nebulizer reservoir (i.e., in the cup and recycled again for nebulization) due to the presence of a baffle system, and only smaller droplets (i.e., the secondary aerosol) can escape the baffle system and are ready for inhalation by the patient. Modern ultrasonic nebulizers are designed to contain a speed-control fan within the device that aids in directing the aerosol droplets via mouthpiece for inhalation. These nebulizers can be activated at a frequency (i.e., above 1 MHz), hence producing aerosols with an MMAD of 2–12 μm, which means that the output of these advanced nebulizers is two to three times better than that of air-jet nebulizers.

Ultrasonic nebulizers are divided into two categories based on the mechanism of aerosol production: capillary-wave formation and cavitational-bubble formation (Fig. 1.14). In the capillary-wave formation mechanism, high-frequency vibrations (generated by piezoelectric crystals) are used to produce a capillary jet, which breaks and generates aerosol droplets. There is a general indirect relationship between the frequency

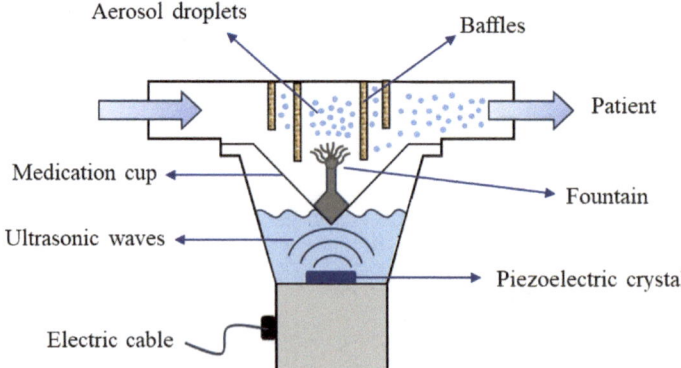

**Fig. 1.13** The general design of an ultrasonic nebulizer is distinguished by a piezoelectric crystal at the base of the nebulizer reservoir which generates vibrations in the aqueous formulation, causing the formation of inhalable aerosol droplets

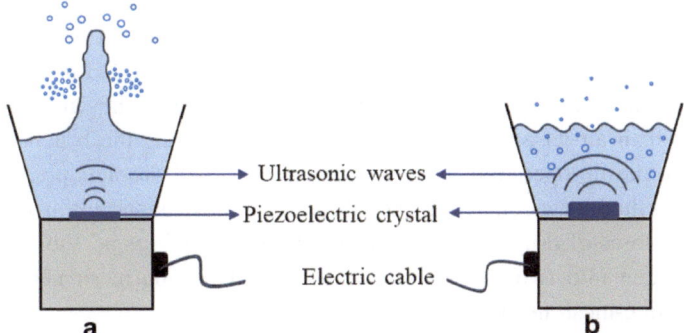

**Fig. 1.14** The mechanisms of generating aerosol droplets via two main categories of ultrasonic nebulizers: **a** capillary-wave formation, and **b** cavitational-bubble formation

of vibrations and the size of the droplets generated; the higher the vibrations, the smaller the droplets (Avvaru et al. 2006). In cavitational-bubble formation, lower-frequency vibrations are used when compared to capillary-wave nebulizers. In this mechanism, the lower frequency generates bubbles inside the aqueous formulation. Upon reaching the surface of the liquid formulation, these bubbles burst when their internal pressure become equal to their external pressure, therefore causing small droplets to form for inhalation (Khan et al. 2013).

Ultrasonic nebulizers may be associated with reduced popularity, which may be accounted for by a handful of key reasons. For example, ultrasonic nebulizers require a power source; they can also damage sensitive drug molecules (i.e., heat-sensitive drugs and proteins), which is a direct result of heat generated during nebulization, causing an increase in the temperature of the nebulizer fluid by approximately 10–15 °C after a few minutes of nebulization. Furthermore, these nebulizers are not strongly recommended for nebulizing viscus solutions (e.g., antibiotics) or suspensions (e.g., corticosteroids). Moreover, after complete nebulization, the performance of ultrasonic nebulizers (e.g., output) is generally lower than that of air-jet nebulizers (Taylor and Hoare 1993; Khan et al. 2021a).

### 1.2.3.3   Vibrating Mesh Nebulizers

These nebulizers were introduced in the early 1920s. As indicated, vibrating-mesh nebulizers mainly consist of a thin metallic plate with micronized tapered holes (perforated by laser-drilled technology). These nebulizers also contain piezoelectric crystals, but they do not produce aerosols directly, as they do in the case of ultrasonic nebulizers. The piezoelectric crystal essentially vibrates the plate that in turn generates aerosol droplets. It is noteworthy that one side of the plate is in direct contact with the formulation to be aerosolized, where the piezoelectric crystal vibrations push the formulation through the micro-sized holes; thus, from the other side of the plate, a mist of small slow-moving droplets are generated for inhalation. The output rate is directly related to the diameters of the holes. Generally, these nebulizers take longer to nebulize, but their residual volume is very small (lower formulation loss) as compared to air-jet and ultrasonic nebulizers (Dhand 2002; Khan et al. 2021b). Mesh nebulizers can be operated using an alternating current (AC) electric power supply or via a battery. Furthermore, they are much lighter and more portable than air-jet and ultrasonic nebulizers. Vibrating-mesh nebulizers are also relatively silent during operation, and the generated aerosol droplets possess a narrower size distribution than their counterpart nebulizers (i.e., air-jets and ultrasonic nebulizers); therefore, high doses of drug can be deposited in the "deep" lungs. Vibrating-mesh nebulizers are divided into two categories: (1) actively vibrating-mesh nebulizers and (2) passively vibrating-mesh nebulizers.

### Actively Vibrating Mesh Nebulizers

Aeroneb Go, Aeroneb Pro (Aerogen Inc., Galway, Ireland), and Pari e-Flow (Pari Respiratory Equipment, Inc., Midlothian, United States) are examples of actively vibrating-mesh nebulizers (Fig. 1.15). These nebulizers have a dome-shaped covering containing a metallic perforated plate with an aperture size of 4 $\mu$m (Aerogen nebulizers) and 20 $\mu$m (Pari e-Flow nebulizer). These perforated plates are directly attached to a piezoelectric crystal vibrational element. For example, in the Aerogen nebulizers, upon nebulization (when an electric current is provided), a "micropump" technology starts to generate aerosol droplets from aqueous formulations via upward and downward movements of the perforated plate (Hess et al. 2011; Khan et al. 2013). The size and distribution of the generated aerosol droplets are controlled by the size of the apertures in the plate. Aeroneb Pro slightly increases the temperature (i.e., circa 3 °C) of the formulation after 5 min of nebulization (Aerogen 2018). This temperature change depends upon the formulation volume in the nebulizer reservoir and nebulizer type (similarly in other active vibrating mesh nebulizers). Changes in the drug concentration in the nebulizer reservoir are significantly lower as compared to the air-jet and ultrasonic nebulizers. Vibrating-mesh nebulizers leave very little dead volume upon completed nebulization and deposit higher fine particle fractions of aerosol droplets (Kuo et al. 2019).

### Passively Vibrating Mesh Nebulizers

Omron MicroAir NE-U22 (Omron Healthcare, Inc., Kyoto, Japan) and I-neb Adaptive Aerosol Delivery (AAD) (Respironics, United States) nebulizers are examples of passively vibrating-mesh nebulizers (Fig. 1.16). These nebulizers contain piezoelectric crystals that produce vibrations when an electric current is applied. These vibrations are transmitted via a transducer horn that is in contact with the aqueous formulation (Elhissi et al. 2007). The vibrations then create waves in the formulation (located in the nebulizer reservoir) and move towards the perforated plate (Omron MicroAir contains about 6000 holes with a 3 $\mu$m diameter), causing it to vibrate and extrude the liquid for generating aerosols. The silent and passive movement of the mesh nebulizer generates smaller aerosol droplets and produces higher aerosol outputs (Waldrep and Dhand 2008; Newman and Gee-Turner

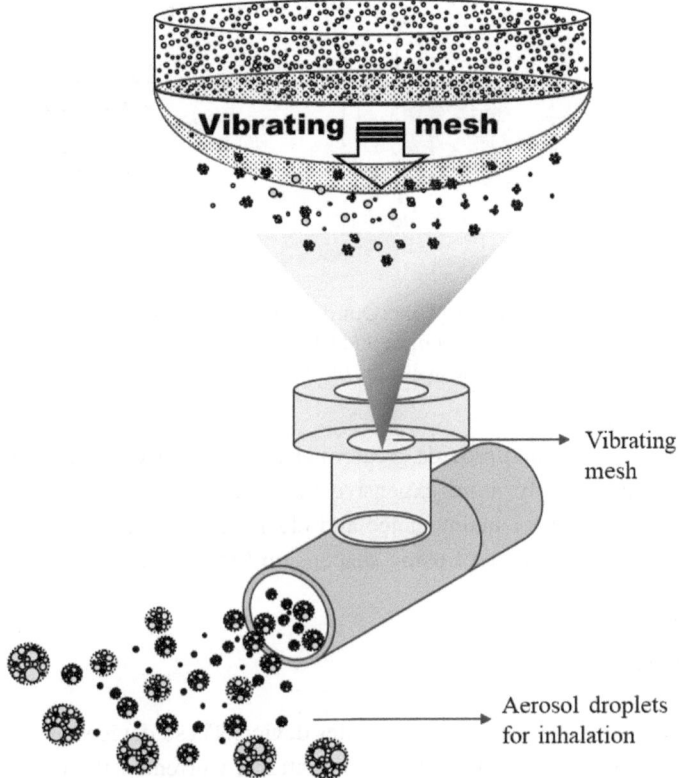

**Fig. 1.15** A schematic diagram of the Aeroneb Pro vibrating-mesh nebulizer, illustrating the mechanism by which the nebulizer generates aerosols from the aqueous formulation

2005). The AAD device is also considered a "smart" nebulizer that contains a breath-actuated system in order to control the inhaled dose. This can be achieved by monitoring the patient's breathing pattern over successive tidal breaths to inhale the drug content (i.e., 50–80%) with each inspiration until the pre-programmed dose is inhaled (Denyer et al. 2004).

Both actively and passively vibrating mesh nebulizers are greatly dependent upon the characteristics of the formulation used for nebulization. Both of these main categories of nebulizers generate continuous aerosols from aqueous formulations, even if they possess higher viscosities (i.e., more than 2 cP), and such physicochemical properties do not affect the formulation output (Ghazanfari et al. 2007; Carvalho and McConville 2016). Employing these formulations with higher viscosity may generate aerosols with smaller droplet sizes. The impact of surface tension on the properties of aerosols is not very clear; however, it is established that formulations with lower viscosities and lower surface

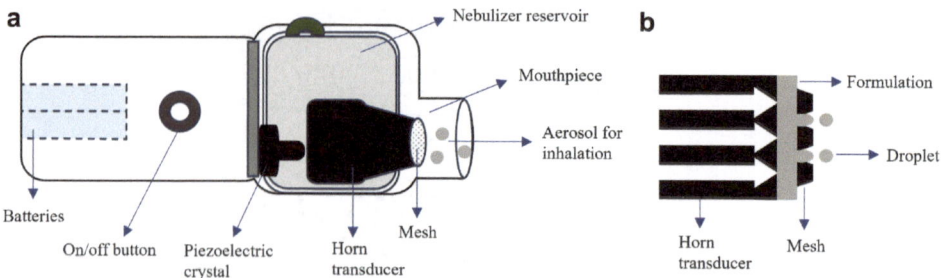

**Fig. 1.16** Diagrammatic presentation of the Omron MicroAir NE-U22 nebulizer, showing **a** the general design of the vibrating mesh nebulizer, and **b** the formation of aerosol droplets from the mesh upon extrusion of the fluid through the apertures of the mesh

tensions are advantageous and provide greater nebulization performance. These vibrating-mesh nebulizers are slightly more expensive than other types of nebulizers. All mesh nebulizers involve a need for maintenance and cleaning after each use to avoid aperture blockage (especially in the case of using suspension formulations).

### 1.2.4  Soft Mist Inhalers

Soft mist inhalers are a new category of inhaler device. These devices are propellant-free, multidose, hand-held and contain liquid formulations. Currently, the only commercially available soft mist inhaler is Respimat®, which generates an aerosol cloud with a relatively high fine particle fraction as compared to traditional pMDIs and DPIs. The Respimat® was developed to overcome the drawbacks of counterpart inhaler devices to effectively generate and deliver aerosol droplets from solutions. The Respimat® does not require a power source, batteries or a spacer. The speed of the generated aerosol spray which exits from the Respimat® device is comparatively slower and of longer duration than pMDIs, and therefore requires minimal co-ordination of actuation and the patient's inhalation (Hochrainer et al. 2005). This results in a decrease in oropharyngeal and increase lung drug deposition, allowing for the administration of medications at lower nominal dosages without compromising efficacy (Anderson 2006).

The Respimat® contains a hinged cap similar to other hand-held inhalers such as DPIs and pMDIs (Dalby et al. 2004). The drug cartridge (stores medication as a solution), an aluminium cylinder with a plastic, double-walled collapsible bag, that releases the formulation solution when the bag contracts (Fig. 1.17). Either water or ethanol can be used to make the solution, with benzalkonium chloride and ethylene diamine tetra-acetic acid (EDTA) added as preservatives. Each puff contains a very low concentration of preservative: roughly 0.44 µg of benzalkonium chloride and 2.2 µg of EDTA.

**Fig. 1.17** A schematic diagram of Respimat® as a soft mist inhaler illustrating its key components

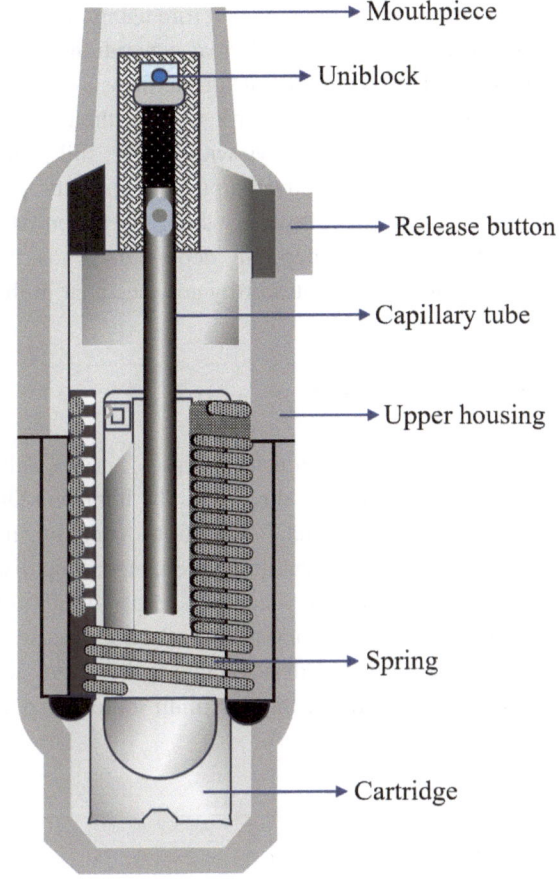

Mouthpiece

Uniblock

Release button

Capillary tube

Upper housing

Spring

Cartridge

   The first commercially available product has a dose indicator (with a total of 120 actuations) with a color-coded-scheme. This indicator does not count the individual doses but provides an estimate of the remaining number of doses. After 120 doses, a locking mechanism is established which prevents the base from rotating and hence preventing additional doses from being actuated. In the Respimat®, by rotating the device's base 180°, a spring is compressed, generating mechanical force to aerosolize the drug dosage. Additionally, a measured quantity of medicine (in general 10–15 µl) is transferred from the drug cartridge to the pump cylinder via a capillary tube (Dalby et al. 2004). The medicine is forced through a nozzle system known as the "uniblock" by the energy from the compressed spring when the dose-release button is activated (Fig. 1.17). The uniblock is around $2 \times 2.5$ mm and is made of a silicon wafer bonded to a glass plate. The generated aerosols from the Respimat® contain a higher fine particle fraction, demonstrating higher emitted dose deposition in the lungs and lower into the oropharynx when compared to DPIs and pMDIs (Zierenberg 1999).

It is important to know that before using the inhaler for the first time, the device should be primed by actuating until clear generation of a cloud appears, followed by three more actuations. This is significant for the measurement of spray content uniformity to achieve 100% of target volume. After a week without usage, priming with one dose of actuation is advised, and after 21 days without use, full priming is advised. The release button for inhalation should be activated/pushed when the patient takes a deep and slow breath in through the mouth, following which the patient should hold their breath for 10 s which is recommended for deep lung deposition. Thus, the development of the Respimat® as a soft mist inhaler is a noteworthy advancement in the administration of drugs to the lungs.

## 1.3    Conclusion

In the treatment of pulmonary diseases, aerosol delivery is still the primary option for the delivery of formulations directly to the lungs via DPIs, pMDIs, or nebulizers. It is important to acknowledge that formulation deposition in the lungs depends on many factors, such as the patient's breathing pattern, age, lung diseases, anatomy of the airways, inhalation techniques and coordination, and use of add-on devices. Similarly, it is not universal that one device is appropriate for all lung diseases. All drug delivery devices have associated advantages and disadvantages; therefore, adequate education and training should be provided to achieve the maximum output and effectively counter/manage pulmonary diseases.

## References

Aerogen. 2018. Aerogen Pro System Instruction Manual. *In:* AEROGEN.COM (ed.). Ireland: Aerogen Ltd., pp: 1–40

Almurshedi, A. S., Aljunaidel, H. A., Alquadeib, B., Aldosari, B. N., Alfagih, I. M., Almarshidy, S. S., Eltahir, E. K. D. & Mohamoud, A. Z. 2021. Development of Inhalable Nanostructured Lipid Carriers for Ciprofloxacin for Noncystic Fibrosis Bronchiectasis Treatment. *Int J Nanomedicine*, 16, 2405–2417

Anderson, P. 2006. Use of Respimat Soft Mist inhaler in COPD patients. *Int J Chron Obstruct Pulmon Dis*, 1, 251–259

Avvaru, B., Patil, M. N., Gogate, P. R. & Pandit, A. B. 2006. Ultrasonic atomization: Effect of liquid phase properties. *Ultrasonics*, 44, 146–158

Benedini, L. & Messina, P. 2022. Chapter 8 - Proniosomes and niosomes for enhanced drug delivery. *In:* Nayak, A. K., Hasnain, M. S., Aminabhavi, T. M. & Torchilin, V. P. (eds.) *Systems of Nanovesicular Drug Delivery*. Academic Press

Capanoglu, M., Dibek Misirlioglu, E., Toyran, M., Civelek, E. & Kocabas, C. N. 2015. Evaluation of inhaler technique, adherence to therapy and their effect on disease control among children with asthma using metered dose or dry powder inhalers. *Journal of Asthma*, 52, 838–845

Carvalho, T. C. & Mcconville, J. T. 2016. The function and performance of aqueous aerosol devices for inhalation therapy. *Journal of Pharmacy and Pharmacology*, 68, 556–578

D'Angelo, I., Conte, C., Miro, A., Quaglia, F. & Ungaro, F. 2015. Pulmonary drug delivery: a role for polymeric nanoparticles? *Curr Top Med Chem*, 15, 386–400

Dalby, R., Spallek, M. & Voshaar, T. 2004. A review of the development of Respimat® Soft Mist™ Inhaler. *International Journal of Pharmaceutics*, 283, 1–9

Denyer, J., Nikander, K. & Smith, N. 2004. Adaptive aerosol delivery (AAD®) technology. *Expert Opinion on Drug Delivery*, 1, 165–176

Dhand, R. 2002. Nebulizers that use a vibrating mesh or plate with multiple apertures to generate aerosol. *Respir Care*, 47, 1406–1416

Dhand, R. 2003. New Nebuliser Technology - Aerosol Generation by using a Vibrating Mesh or Plate with Multiple Apertures. (2003), University of Missouri Hospital and Clinics, for Omeron Healthcare, Inc.

Dolovich, M. B. & Dhand, R. 2011. Aerosol drug delivery: developments in device design and clinical use. *Lancet*, 377, 1032–1045

Elhissi, A. 2017. Liposomes for Pulmonary Drug Delivery: The Role of Formulation and Inhalation Device Design. *Curr Pharm Des*, 23, 362–372

Elhissi, A. & Taylor, K. M. G. 2005. Delivery of liposomes generated from pro liposomes using air-jet, ultrasonic and vibrating-mesh nebulisers. *J. Drug Del. Sci.Technol.*, 15, 261–265

Elhissi, A. M. A., Faizi, M., Naji, W. F., Gill, H. S. & Taylor, K. M. G. 2007. Physical stability and aerosol properties of liposomes delivered using an air-jet nebulizer and a novel micropump device with large mesh apertures. *International Journal of Pharmaceutics*, 334, 62–70

Elhissi, A., Hidayat, K., Phoenix, D. A., Mwesigwa, E., Crean, S., Ahmed, W., Faheem, A. & Taylor, K. M. G. 2013. Air-jet and vibrating-mesh nebulization of niosomes generated using a particulate-based proniosome technology. *International Journal of Pharmaceutics*, 444, 193–199

Errington, R. 2012. Inhalation manufacturing: cold fill, pressure fill, and finding the right partner. *ONdrug Delivery*, 37, 30–32

Ezzati Nazhad Dolatabadi, J., Hamishehkar, H. & Valizadeh, H. 2015. Development of dry powder inhaler formulation loaded with alendronate solid lipid nanoparticles: solid-state characterization and aerosol dispersion performance. *Drug Development and Industrial Pharmacy*, 41, 1431–1437

Fink, J. B. 2000. Metered-dose inhalers, dry powder inhalers, and transitions. *Respiratory Care*, 45, 623

Ghazanfari, T., Elhissi, A. M. A., Ding, Z. & Taylor, K. M. G. 2007. The influence of fluid physicochemical properties on vibrating-mesh nebulization. *International Journal of Pharmaceutics*, 339, 103–111

Goyal, A. K., Garg, T., Bhandari, S. & Rath, G. 2017. Chapter 22 - Advancement in pulmonary drug delivery systems for treatment of tuberculosis. *In:* Andronescu, E. & Grumezescu, A. M. (eds.) *Nanostructures for Drug Delivery*. Elsevier

Grossman, J. 1994. The evolution of inhaler technology. *J Asthma*, 31, 55–64

Hamilton, M., Leggett, R., Pang, C., Charles, S., Gillett, B. & Prime, D. 2015. In vitro dosing performance of the ELLIPTA® dry powder inhaler using asthma and COPD patient inhalation profiles replicated with the electronic lung (eLung™). *Journal of Aerosol Medicine and Pulmonary Drug Delivery*, 28, 498–506

Haswell, S. 2020. Cold filling versus pressure filling: the case for versatile, fully integrated CDMOs. *ONdrugDelivery*. Lewes, UK: Frederick Furness Publishing Ltd.

Hecker, M., Linder, T., Ott, J., Walmrath, H.-D., Lohmeyer, J., Vadász, I., Marsh, L. M., Herold, S., Reichert, M., Buchbinder, A., Morty, R. E., Bausch, B., Fischer, T., Schulz, R., Grimminger, F., Witzenrath, M., Barnes, M., Seeger, W. & Mayer, K. 2015. Immunomodulation by lipid emulsions in pulmonary inflammation: a randomized controlled trial. *Critical Care*, 19, 226

Hess, R. D. 2000. Nebulizers: Principles and Performance. *Respiratory care*, 45, 609–622

Hess, D., Fisher, D., Williams, P., Pooler, S. & Kacmarek, R. M. 1996. Medication nebulizer performance. Effects of diluent volume, nebulizer flow, and nebulizer brand. *Chest*, 110, 498–505

Hess, R. D., Macintyre, R. N., Mishoe, C. S. & Galvin, F. W. 2011. *Respiratory Care: Principles and Practice*, Jones & Bartlett Learning

Heyder, R. S., Zhong, Q., Bazito, R. C. & Da Rocha, S. R. 2017. Cellular internalization and transport of biodegradable polyester dendrimers on a model of the pulmonary epithelium and their formulation in pressurized metered-dose inhalers. *International Journal of Pharmaceutics*, 520, 181–194

Hochrainer, D., Hölz, H., Kreher, C., Scaffidi, L., Spallek, M. & Wachtel, H. 2005. Comparison of the aerosol velocity and spray duration of Respimat® Soft Mist™ inhaler and pressurized metered dose inhalers. *Journal of Aerosol Medicine*, 18, 273–282

Hugh, D. C. S. 2003. The influence of formulation variables on the performance of alternative propellant-driven metered dose inhalers. *Advanced Drug Delivery Reviews*, 55, 807–828

Kendrick, A. H., Smith, E. C. & Wilson, R. S. 1997. Selecting and using nebuliser equipment. *Thorax*, 52 Suppl 2, S92–S101

Khan, I., Elhissi, A., Shah, M., Alhnan, M. A. & Ahmed, W. 2013. 9 - Liposome-based carrier systems and devices used for pulmonary drug delivery. *In:* Davim, J. P. (ed.) *Biomaterials and Medical Tribology*. Woodhead Publishing

Khan, I., Yousaf, S., Alhnan, M. A., Ahmed, W., Elhissi, A. & Jackson, M. J. 2016. Design Characteristics of Inhaler Devices Used for Pulmonary Delivery of Medical Aerosols. *In:* Ahmed, W. & Jackson, M. J. (eds.) *Surgical Tools and Medical Devices*. Cham: Springer International Publishing, pp: 573–591

Khan, I., Yousaf, S., Subramanian, S., Albed Alhnan, M., Ahmed, W. & Elhissi, A. 2018. Proliposome tablets manufactured using a slurry-driven lipid-enriched powders: Development, characterization and stability evaluation. *International Journal of Pharmaceutics*, 538, 250–262

Khan, I., Apostolou, M., Bnyan, R., Houacine, C., Elhissi, A. & Yousaf, S. S. 2020a. Paclitaxel-loaded micro or nano transfersome formulation into novel tablets for pulmonary drug delivery via nebulization. *International Journal of Pharmaceutics*, 575, 118919

Khan, I., Lau, K., Bnyan, R., Houacine, C., Roberts, M., Isreb, A., Elhissi, A. & Yousaf, S. 2020b. A Facile and Novel Approach to Manufacture Paclitaxel-Loaded Proliposome Tablet Formulations of Micro or Nano Vesicles for Nebulization. *Pharmaceutical Research*, 37, 116

Khan, I., Hussein, S., Houacine, C., Khan Sadozai, S., Islam, Y., Bnyan, R., Elhissi, A. & Yousaf, S. 2021a. Fabrication, characterization and optimization of nanostructured lipid carrier formulations using Beclomethasone dipropionate for pulmonary drug delivery via medical nebulizers. *International Journal of Pharmaceutics*, 598, 120376

Khan, I., Needham, R., Yousaf, S., Houacine, C., Islam, Y., Bnyan, R., Sadozai, S. K., Elrayess, M. A. & Elhissi, A. 2021b. Impact of phospholipids, surfactants and cholesterol selection on the performance of transfersomes vesicles using medical nebulizers for pulmonary drug delivery. *Journal of Drug Delivery Science and Technology*, 66, 102822

Khan, I., Al-Hasani, A., Khan, M. H., Khan, A. N., Alam, F. E., Sadozai, S. K., Elhissi, A., Khan, J. & Yousaf, S. 2023. Impact of dispersion media and carrier type on spray-dried proliposome powder formulations loaded with beclomethasone dipropionate for their pulmonary drug delivery via a next generation impactor. *PLoS One*, 18, e0281860

Kuo, Y.-M., Chan, W.-H., Lin, C.-W., Huang, S.-H. & Chen, C.-C. 2019. Characterization of Vibrating Mesh Aerosol Generators. *Aerosol and Air Quality Research*, 19, 1678–1687

Labiris, N. R. & Dolovich, M. B. 2003. Pulmonary drug delivery. Part II: the role of inhalant delivery devices and drug formulations in therapeutic effectiveness of aerosolized medications. *Br J Clin Pharmacol*, 56, 600–612

Laube, B. L. & Dolovich, M. B. 2014. 66 - Aerosols and Aerosol Drug Delivery Systems. *In:* Adkinson, N. F., Bochner, B. S., Burks, A. W., Busse, W. W., Holgate, S. T., Lemanske, R. F. & O'Hehir, R. E. (eds.) *Middleton's Allergy (Eighth Edition)*. London: W.B. Saunders

Leong, E. W. X. & Ge, R. 2022. Lipid Nanoparticles as Delivery Vehicles for Inhaled Therapeutics. *Biomedicines*, 10

Liang, Z., Ni, R., Zhou, J. & Mao, S. 2015. Recent advances in controlled pulmonary drug delivery. *Drug Discovery Today*, 20, 380–389

Malamatari, M., Charisi, A., Malamataris, S., Kachrimanis, K. & Nikolakakis, I. 2020. Spray Drying for the Preparation of Nanoparticle-Based Drug Formulations as Dry Powders for Inhalation. *Processes*, 8, 788

Mccallion, O. N. M., Taylor, K. M. G., Bridges, P. A., Thomas, M. & Taylor, A. J. 1996. Jet nebulisers for pulmonary drug delivery. *International Journal of Pharmaceutics*, 130, 1–11

Newman, S. P. 2005. Principles of metered-dose inhaler design. *Respir Care*, 50, 1177–1190

Newman, S. P. & Gee-Turner, A. 2005. The Omron MicroAir vibrating mesh technology nebuliser, a 21st century approach to inhalation therapy. *Journal of Applied Therapeutic Research* 5, 29–33

NHS. 2024. Respiratory disease. Accessed date 15th March 2024. https://www.england.nhs.uk/our work/clinical-policy/respiratory-disease/

Nicolini, G., Cremonesi, G. & Melani, A. S. 2010. Inhaled corticosteroid therapy with nebulized beclometasone dipropionate. *Pulmonary Pharmacology & Therapeutics*, 23, 145–155

O'Callaghan, C. & Barry, P. W. 1997. The science of nebulised drug delivery. *Thorax*, 52 Suppl 2, S31–S44

Orizondo, R. A., Fabiilli, M. L., Morales, M. A. & Cook, K. E. 2016. Effects of Emulsion Composition on Pulmonary Tobramycin Delivery During Antibacterial Perfluorocarbon Ventilation. *J Aerosol Med Pulm Drug Deliv*, 29, 251–259

Pindiprolu, S., Kumar, C. S. P., Kumar Golla, V. S., Likitha, P., Shreyas Chandra, K., Esub Basha, S. K. & Ramachandra, R. K. 2020. Pulmonary delivery of nanostructured lipid carriers for effective repurposing of salinomycin as an antiviral agent. *Med Hypotheses*, 143, 109858

Price, D., Roche, N., Virchow, J. C., Burden, A., Ali, M., Chisholm, A., Lee, A. J., Hillyer, E. V. & Von Ziegenweidt, J. 2011. Device type and real-world effectiveness of asthma combination therapy: an observational study. *Respiratory Medicine*, 105, 1457–1466

Rubin, B. K. & Fink, J. B. 2005. Optimizing aerosol delivery by pressurized metered-dose inhalers. *Respir Care*, 50, 1191–1200

Smyth, H., Hickey, A. J., Brace, G., Barbour, T., Gallion, J. & Grove, J. 2006. Spray pattern analysis for metered dose inhalers I: Orifice size, particle size, and droplet motion correlations. *Drug Dev Ind Pharm*, 32, 1033–1041

Taylor, K. M. G. & Hoare, C. 1993. Ultrasonic nebulisation of pentamidine isethionate. *International Journal of Pharmaceutics*, 98, 45–49

Telko, M. J. & Hickey, A. J. 2005. Dry powder inhaler formulation. *Respir Care*, 50, 1209–1227

Traini, D., Rogueda, P., Young, P. & Price, R. 2005. Surface Energy and Interparticle Force Correlation in Model pMDI Formulations. *Pharmaceutical Research*, 22, 816–825

Traini, D., Young, P. M., Rogueda, P. & Price, R. 2006. The Use of AFM and Surface Energy Measurements to Investigate Drug-Canister Material Interactions in a Model Pressurized Metered Dose Inhaler Formulation. *Aerosol Science and Technology*, 40, 227–236

Vallorz, E., Sheth, P. & Myrdal, P. 2019. Pressurized metered dose inhaler technology: manufacturing. *AAPS PharmSciTech*, 20, 177

Waldrep, J. C. & Dhand, R. 2008. Advanced nebulizer designs employing vibrating mesh/aperture plate technologies for aerosol generation. *Curr Drug Deliv*, 5, 114–119

WHO. 2024. WHO mortality database, interactive platform visualizing mortality data. Access date, 13th March 2024. https://platform.who.int/mortality/themes/theme-details/topics/topic-det ails/MDB/respiratory-diseases. *Respiratory diseases*

Young, P. M., Price, R., Lewis, D., Edge, S. & Traini, D. 2003. Under pressure: predicting pressurized metered dose inhaler interactions using the atomic force microscope. *Journal of Colloid and Interface Science*, 262, 298–302

Zierenberg, B. 1999. Optimizing the in vitro performance of Respimat. *Journal of Aerosol Medicine*, 12, S-19–S-24

# Manufacturing Strategies for Liposome and Proliposome-Based Drug Delivery Systems

## Abstract

In this chapter, various methods of preparing liposomes were reviewed with respect to the desired liposome characteristics such as size and number of bilayers. Thin-film hydration technique is the universally recognized method for preparing multilamellar liposome vesicles (MLVs), while subsequent sonication or extrusion can be used to generate small unilamellar liposome vesicles (SUVs) or large unilamellar liposome vesicles (LUVs), respectively. Liposomes made using the traditional techniques (e.g. thin-film hydration) have a range of instabilities such as phospholipid hydrolysis and oxidation, with subsequent vesicle aggregation and loss of the originally entrapped material from the liposomes. This is further complicated by the possible microbial contamination of the liposome formulations. Freeze-drying has been suggested as a method to stabilize liposomes through producing a lyophilized powder of the vesicles. Unfortunately, the stressful effect of freeze-drying may damage the liposome structures, causing them to aggregate or fuse during rehydration with concomitant loss of the originally entrapped material. Cryoprotectant materials (e.g. carbohydrates) have been included in the liposome formulations prior to freeze-drying in order to protect the liposomes from the deleterious effects of freeze-drying. As an alternative to freeze-drying, proliposome technologies are stable phospholipid formulations that are either particulate-based or solvent-based, and can generate liposomes upon addition of water just before administration. As for traditional liposomes, vesicles generated from proliposomes can be further processed to generate nano-sized vesicles (e.g. by probe-sonication). Liposomes generated from proliposomes can be produced at a larger scale using methods like high-pressure homogenization. An approach employing fluid-bed coating to manufacture proliposome granules consisting of sugar coated with lipids and steroid, followed by hydration and high-pressure homogenization to generate SUVs, has been introduced in our laboratory. This was followed by freeze-drying of the self-cryoprotected SUVs to generate stable lyophilized formulations offering high drug entrapment efficiency. Overall, further work on scaling up liposomes generated from proliposomes is needed to unlock the full potential of proliposome technologies for developing novel inhalable formulations.

## 2.1    Introduction

Proliposomes have emerged as a transformative technology in drug delivery, offering a practical solution to several challenges associated with storage, stability, and large-scale production of traditional liposomes (Choudhary et al. 2022; Dhiman et al. 2022; Elhissi 2017). Proliposome drug delivery systems spontaneously convert into liposomes upon hydration, ensuring easier handling, enhanced storage stability, and greater potential for manufacturing on a large scale compared to traditional liposomes, such as those made using the thin-film hydration method (Choudhary et al. 2022; Dhiman et al. 2022; Elhissi 2017). Unlike traditional liposome formulations, which often require complex storage conditions and suffer from degradation over time, proliposomes provide a more robust and scalable platform for manufacturing liposomes (Choudhary et al. 2022; Dhiman et al. 2022; Elhissi 2017).

The primary focus of this chapter is on proliposome technologies as alternatives to the traditional methods for generating liposomes. Key considerations for optimizing proliposome formulations are discussed, including the role of carrier, solvent, lipid composition, and process parameters.

## 2.2    Liposomes: Definition, Structure, and Formation

### 2.2.1    Definition and Characteristics of Liposomes

Liposomes are vesicles formed through the self-assembly of diacyl-chain phospholipids, creating multiple lipid bilayers in the form of vesicles within aqueous solutions (Akbarzadeh et al. 2013; Naeem 2017; Nsairat et al. 2022). Their name originates from the Greek terms "lipo" (fat) and "soma" (body), reflecting their lipid composition (Naeem 2017). The phospholipid bilayer membrane has a hydrophobic tail and a hydrophilic head, resulting in an amphiphilic structure that is suitable for encapsulating a wide range of therapeutic molecules, with various physicochemical characteristics (Akbarzadeh et al. 2013; Naeem 2017; Nsairat et al. 2022). Liposomes can be composed of either natural or synthetic phospholipids, and their lipid composition significantly influences their characteristics, including vesicle particle size and bilayer (membrane) rigidity, and overall formulation stability (Akbarzadeh et al. 2013; Naeem 2017; Nsairat et al. 2022). In drug delivery, liposomes offer significant advantages, such as targeted delivery, enhanced drug bioavailability, and controlled release profile of the entrapped drug, aiming to minimize side effects and improve therapeutic outcomes (Naeem 2017; Sercombe et al. 2015).

## 2.3 Classification of Liposomes and Their Methods of Preparation

Liposomes are highly versatile vesicular structures employed in drug delivery systems, classified primarily by size and lamellarity (i.e., the number of bilayers). The main types include small unilamellar vesicles (SUVs), large unilamellar vesicles (LUVs), multilamellar vesicles (MLVs), oligolamellar vesicles (OLVs), and multivesicular vesicles (MVLs). Each type exhibits distinct characteristics based on its structure and preparation method, making them suitable for specific applications. Various methods have been developed for preparing the different liposomal types, including thin-film hydration, ethanol injection, and double emulsion methods, which are discussed in further details within the subsequent sections, and are also outlined in Table 2.1. A variety of techniques is employed to achieve optimal drug loading and maintain liposome stability during storage. Mechanical approaches such as sonication, extrusion, and high-pressure homogenization facilitate the production of liposomes with defined sizes (e.g. in the nanoscale) and structural characteristics. Freeze-drying (lyophilization) is also crucial for extending the shelf life of liposomes and ensuring their stability, often with the aid of cryoprotectants (e.g. carbohydrates) in appropriate concentrations.

### 2.3.1 Multilamellar Vesicles (MLVs)

Multilamellar liposome vesicles (MLVs) vary in size and consist of more than five concentric lipid bilayers (Andra et al. 2022; Lombardo and Kiselev 2022; Zhang and Sun 2021). MLVs are straightforward to produce, usually by hydrating dry lipid films; however, their heterogeneous size distribution can sometimes be a serious limitation for application in drug delivery (Andra et al. 2022; Lombardo and Kiselev 2022; Zhang and Sun 2021). MLVs, because of their multiple bilayers, offer robust protection for the encapsulated material and allow for extended drug release (Andra et al. 2022; Lombardo and Kiselev 2022; Zhang and Sun 2021). MLVs are usually prepared using the thin-film hydration, reverse-evaporation, or detergent removal methods (Lombardo and Kiselev 2022).

#### 2.3.1.1 Thin-Film Hydration Method

Thin-film hydration method, also referred to as "Bangham method", is one of the most commonly and the first reported method utilized for preparing MLVs. It involves dissolving lipids in an organic solvent (commonly a mixture of chloroform and methanol) within

**Table 2.1** Summary of traditional liposome preparation methods

| Preparation method | Liposome types | Principle | General procedure | Advantages | Limitations |
|---|---|---|---|---|---|
| Thin-film hydration | MLVs, OLVs, GUVs | Lipid film formation and hydration to form MLVs, OLVs, or GUVs | Dissolve lipids in organic solvent, evaporate solvent to form a thin film, hydrate with aqueous solution | Simple, widely used, suitable for various drugs | Requires high energy, heterogeneous size distribution |
| Reverse-phase evaporation | MLVs, LUVs | Forms MLVs or LUVs by creating a water-in-oil emulsion, which collapses into vesicles upon solvent removal | Dissolve lipids in organic solvent, emulsify with aqueous phase, remove solvent under vacuum | High encapsulation efficiency, ideal for encapsulation of hydrophilic drugs | Residual solvents may affect stability, limited scaling-up |
| Detergent removal (depletion) | MLVs, LUVs | Uses detergent-solubilized lipids to form mixed micelles that transition into liposomes as detergent is gradually removed | Mix lipids with detergent, remove detergent gradually, dialysis, or resin adsorption | High reproducibility, good for protein incorporation | Detergent residues, low entrapment of hydrophobic drugs |
| Ethanol injection | SUVs | Rapid injection of lipids dissolved in ethanol into an aqueous phase to form small SUVs | Inject ethanol-lipid solution into heated buffer, allow self-assembly of liposomes | Simple, reproducible, nontoxic solvent (ethanol) is used | Residual ethanol, heterogeneous vesicle sizes |
| Ether injection | SUVs, LUVs | Similar to ethanol injection but uses ether, which evaporates upon injection, forming LUVs or SUVs depending on conditions | Inject ether-lipid solution into buffer, heat to evaporate ether | Higher liposome concentration, efficient solvent removal | Exposure to organic solvent, high-temperature risks |

(continued)

**Table 2.1** (continued)

| Preparation method | Liposome types | Principle | General procedure | Advantages | Limitations |
| --- | --- | --- | --- | --- | --- |
| Double emulsification | MVVs | Formation of water-in-oil-in-water emulsion to encapsulate hydrophilic drugs within aqueous cores of vesicles | Create a primary water-in-oil emulsion, emulsify with aqueous phase to form water-in-oil-in-water emulsion, remove solvent | Enhanced hydrophilic drug encapsulation | Complex process, may require high-energy input for stable vesicle formation |

a round-bottomed flask followed by evaporation of the organic solvent(s) under reduced pressure using rotary evaporators. This results in the formation of a thin film onto the inner walls of the flask, which is then hydrated with an aqueous solution, typically under vortex mixing, to generate "milky" dispersions of MLVs (Liu et al. 2022; Lombardo and Kiselev 2022; Lu and Qi 2021). This method is highly adaptable and suitable for various drugs, although it requires significant energy input and may result in vesicles with very large (e.g. 20 μm) and heterogeneous sizes (Liu et al. 2022; Lombardo and Kiselev 2022; Lu and Qi 2021). Additionally, this method presents challenges for large-scale production, faces stability issues during storage of the hydrated vesicles, demonstrates limited entrapment efficiency for hydrophilic drugs, and is time-consuming (Lombardo and Kiselev 2022).

### 2.3.1.2  Reverse-Phase Evaporation Method

Reverse-phase evaporation method offers high encapsulation efficiencies, especially for hydrophilic drugs (Lombardo and Kiselev 2022; Lu and Qi 2021). This method involves creating a water-in-oil emulsion and then removing the organic solvent to form lipid vesicles. The addition of buffer facilitates vesicle formation (Lombardo and Kiselev 2022; Lu and Qi 2021). However, solvent residues can affect stability; thus, careful removal of the organic solvent is needed to ensure biocompatibility and avoid potential toxicities (Lombardo and Kiselev 2022; Lu and Qi 2021).

### 2.3.1.3  Detergent Removal

Detergent removal uses detergents to solubilize lipids, forming mixed micelles that transition into vesicles upon detergent removal via dialysis or resin adsorption (Lombardo and Kiselev 2022). This approach is reproducible and allows efficient incorporation of proteins; however, it may leave detergent residues and offer low encapsulation efficiency for hydrophobic drugs (Lombardo and Kiselev 2022).

### 2.3.2  Small Unilamellar Vesicles (SUVs)

Small unilamellar liposome vesicles (SUVs) are made of a single lipid bilayer, and have diameters between 20 and 100 nm (Andra et al. 2022; de Freitas et al. 2019; Zhong and Zhang 2023). They are created through sonication or high-pressure homogenization of larger vesicles, such as MLVs (Andra et al. 2022; de Freitas et al. 2019; Zhong and Zhang 2023). Their small size and compact structure facilitate their uptake by the cells, making them suitable for targeted drug delivery that can facilitate rapid and efficient cellular uptake (Zhang and Sun 2021). SUVs are usually generated using the solvent injection method followed by sonication (e.g. probe-sonication) (Lombardo and Kiselev 2022).

#### 2.3.2.1   Solvent Injection Method

Solvent injection involves injecting lipids dissolved in ethanol (or ether) into an aqueous buffer, where liposomes spontaneously form upon ethanol dilution (Lombardo and Kiselev 2022). Using ethanol injection presents a simple technique employing a nontoxic solvent, though removing residual ethanol and achieving a uniform size can be a challenge (Lombardo and Kiselev 2022). Ether injection, on the other hand, exposes drugs to higher temperatures and organic solvents, posing potential stability risks (Lombardo and Kiselev 2022).

#### 2.3.2.2   Sonication

Sonication applies mechanical energy via ultrasonic waves to reduce larger vesicles into SUVs (Liu et al. 2022; Lombardo and Kiselev 2022; Lu and Qi 2021). This straightforward method enables the production of smaller vesicles, although high-energy input may degrade sensitive drugs and result in size variability among vesicles. Sonication is commonly performed using titanium-made probe sonicators. Thus, removal of the leaching titanium microparticles from the liposome dispersion is necessary, which is usually achieved via centrifugation to sediment the titanium from the SUVs dispersion (Liu et al. 2022; Lombardo and Kiselev 2022; Lu and Qi 2021).

### 2.3.3   Large Unilamellar Vesicles (LUVs)

Large unilamellar liposome vesicles (LUVs), as their name indicates, are single lipid bilayer vesicles but with larger size than SUVs, with diameters ranging from 100 to 1000 nm (Lombardo and Kiselev 2022; Neves et al. 2009; Zhang and Sun 2021). LUVs provide a larger internal volumes of aqueous phase compared to SUVs, thus, they are appropriate for encapsulating hydrophilic drugs. LUVs are also appropriate carriers of various bioactive agents, including genetic materials (Lombardo and Kiselev 2022; Neves et al. 2009; Zhang and Sun 2021). Their size and structure make them well suited for encapsulating and delivering macromolecules, such as in case of gene therapy (Lombardo and Kiselev 2022; Neves et al. 2009; Zhang and Sun 2021). They are usually prepared using the reverse-phase evaporation or detergent removal methods followed by extrusion through polycarbonate membrane filters, in order to attain the desired particle size (Lombardo and Kiselev 2022).

#### 2.3.3.1   Extrusion

Extrusion is a widely applied technique to produce liposomes of uniform size (Liu et al. 2022; Lombardo and Kiselev 2022; Lu and Qi 2021). By forcing a lipid suspension through polycarbonate membranes of defined pore size, extrusion yields vesicles with a consistent size distribution, which is appropriate for standardized formulations (Liu et al. 2022; Lombardo and Kiselev 2022; Lu and Qi 2021). Despite its advantages, it requires specialized equipment and may be prone to membrane clogging. This is attributed to vesicle characteristics, such as initial size distribution, bilayer rigidity, and possible drug retention by the extruding membrane, sometimes compromising the practicality of this technique (Liu et al. 2022; Lombardo and Kiselev 2022; Lu and Qi 2021).

### 2.3.4  Giant Unilamellar Vesicles (GUVs)

Similar to SUVs and LUVs, Giant unilamellar liposome vesicles (GUVs) are composed of a single lipid bilayer; but are larger in size, typically over 1000 nm in diameter (Andra et al. 2022; Boban et al. 2021; Nair and Bajaj 2023; Zhang and Sun 2021). They are widely used to study membrane biophysical processes, such as phase separation, curvature, and protein interactions, that mimic cellular membrane properties (Andra et al. 2022; Boban et al. 2021; Nair and Bajaj 2023; Zhang and Sun 2021). Owing to their similarity to cellular dimensions, this type of vesicles can be used as model cell systems in studies focused on artificial cell synthesis and in applications where cell-like compartmentalization is essential (Andra et al. 2022; Boban et al. 2021; Nair and Bajaj 2023; Zhang and Sun 2021). Some of the preparation techniques of GUVs include the lipid-film hydration (e.g., electroformation, gel-assisted hydration) and the water-in-oil droplet emulsion technique (Nair and Bajaj 2023).

### 2.3.5  Oligolamellar Vesicles (OLVs)

Oligolamellar liposome vesicles (OLVs) are variable in size and are composed of two to five concentric phospholipid bilayers, i.e. more than unilamellar vesicles but fewer than multilamellar bilayers (Andra et al. 2022; Zhang and Sun 2021). These vesicles offer multiple compartments, which is advantageous in applications requiring controlled, staged release of therapeutic substances (Andra et al. 2022; Zhang and Sun 2021). However, some formulation methods of OLVs require special conditions such as low ionic strength, which can compromise vesicle stability and impose specific formulation requirements (Andra et al. 2022; Zhang and Sun 2021). OLVs are typically produced using the thin-film hydration method followed by extrusion (Koudelka et al. 2010).

### 2.3.6  Multivesicular Vesicles (MVVs)

Multivesicular liposome vesicles (MVVs) are composed of several nonconcentric vesicles with variable sizes that are encapsulated within a single lipid bilayer (Andra et al. 2022; Giuliano et al. 2021; Zhang and Sun 2021). Their preparation involves complex techniques to enclose multiple vesicles within a larger vesicle. The double emulsification method is a common technique for preparing MVVs (Chaurasiya et al. 2022). This technique involves creating a water-in-oil-in-water (W/O/W) emulsion, where an aqueous solution containing the drug is first emulsified into an oil phase, and then the primary emulsion is further emulsified into another aqueous phase (Chaurasiya et al. 2022). The organic solvent is subsequently removed to form the MVVs (Chaurasiya et al. 2022). MVVs are employed in drug delivery systems, aiming for more prolonged drug release and higher formulation stability (Andra et al. 2022; Giuliano et al. 2021; Zhang and Sun 2021). Therefore, they are mainly used in therapeutic applications that demand prolonged dosing, gradual, and consistent release, especially in chronic treatment regimens (Andra et al. 2022; Giuliano et al. 2021; Zhang and Sun 2021). MVVs are typically prepared using the double emulsification technique (Chaurasiya et al. 2022).

### 2.3.6.1 Double Emulsification

The double emulsification technique is commonly used to prepare multivesicular liposomes (MLVs) (Chaurasiya et al. 2022). This process involves creating a water-in-oil-in-water (W/O/W) emulsion, where the primary step emulsifies an aqueous solution containing the drug into an oil phase (Chaurasiya et al. 2022). Following this, the primary emulsion is mixed with a secondary aqueous phase, producing the W/O/W emulsion (Chaurasiya et al. 2022). The final step involves removing the organic solvent, allowing the MVLs to form (Chaurasiya et al. 2022). This method requires strict control over the emulsification parameters to achieve consistent vesicle size and drug encapsulation, thus providing a stable delivery system for sustained drug release applications (Chaurasiya et al. 2022). Producing MVLs remains challenging due to batch-to-batch inconsistencies, extended processing times, residual organic solvents, the need for an efficient sterilization method, and the need to conduct the entire process under an aseptic environment (Chaurasiya et al. 2022).

## 2.4 Challenges and Stability of Liposomes

Liposomes are prone to instability, which can lead to changes in particle size and leakage of their contents. Freeze-drying is commonly used to improve the shelf life and stability of liposomes, but it also introduces physical instabilities to liposome structures, owing to the damaging effects elicited by ice crystals on the vesicles during freezing and vacuum application during sublimation. The following sections outline these challenges in more detail.

### 2.4.1 Storage Stability

To be appropriate for application as a drug delivery system, liposomes must meet strict stability standards, making optimized storage conditions critical for extending their shelf life. Storage stability issues of liposomes include leakage of the originally encapsulated contents, thermosensitivity, and chemical instability (Ball et al. 2016; Sainaga Jyothi et al. 2022). Physical stability is often manifested by aggregation and fusion of the vesicles, leading to particle size enlargement and leakage of the originally entrapped material (Sainaga Jyothi et al. 2022). This phenomenon is influenced by factors such as ionic strength and vesicle-vesicle/vesicle-drug interactions (Sainaga Jyothi et al. 2022). Chemical stability concerns arise primarily from lipid oxidation in presence of oxygen, and hydrolysis in aqueous dispersions, particularly due to the use of unsaturated lipids in the formulation (Sainaga Jyothi et al. 2022). Phospholipid oxidation can be exacerbated by factors such as metal ions and exposure to oxygen, whereas hydrolysis depends on the presence of water molecules, environmental pH, and storage temperature (storage at fridge temperature is desirable for the stability of liposomes) (Ball et al. 2016; Sainaga Jyothi et al. 2022).

## 2.4.2    Freeze-Drying

Freeze-drying, a widely used approach to stabilize many pharmaceutical formulations (e.g. parenterals), can be destabilizing to liposomes, demonstrating the huge significance and complication of the challenge of liposome stability. Owing to the "delicacy" of the liposome structures, freeze-drying involving creation of ice crystals and application of vacuum, cause liposome formulations to "collapse" through massive aggregation/fusion and concomitant loss of the originally entrapped material (Ball et al. 2016; Boafo et al. 2022; Ghanbarzadeh et al. 2013; Sainaga Jyothi et al. 2022). This happens especially when cryoprotectants are absent or insufficient (Ball et al. 2016; Boafo et al. 2022; Ghanbarzadeh et al. 2013; Sainaga Jyothi et al. 2022). Cryoprotective agents such as sugars (e.g., trehalose and sucrose) play a critical role in stabilizing liposomes during freeze-drying (Ball et al. 2016; Boafo et al. 2022; Ghanbarzadeh et al. 2013; Sainaga Jyothi et al. 2022). Appropriate cryoprotective concentrations prevent ice formation within liposomes, maintain particle size, and help preserve the structural integrity upon rehydration (Boafo et al. 2022; Ghanbarzadeh et al. 2013). Freeze-drying can also affect the proportion of drug retained within liposomes (Ball et al. 2016; Boafo et al. 2022; Ghanbarzadeh et al. 2013; Sainaga Jyothi et al. 2022). Thus, although freeze-drying can be successfully employed to stabilize liposomes, the need for involving a cryoprotectant in an appropriate concentration may complicate the formulation and potentially increase the cost of liposome production. Hence, designing convenient alternative methods for generating stable liposomes is highly advantageous. The subsequent sections of this chapter will emphasize proliposomes as highly promising, stable formulations, and feasible technologies for generating liposomes just prior to administration.

## 2.5 Proliposome Technologies and Their Benefits

Proliposomes are categorized into particulate-based and solvent-based types. Particulate-based proliposomes consist of solid carriers that adsorb the drug-lipid mixture, creating a stable powder or granular form (Elhissi et al. 2015). By contrast, solvent-based proliposomes incorporate organic solvents (usually ethanol) to dissolve the lipids, and produce liposomes upon addition of a sufficient amount of the aqueous phase (Elhissi et al. 2015). Proliposomes offer a stable and versatile method for administering liposomes via various routes, including oral, transdermal, pulmonary, parenteral and intranasal (Choudhary et al. 2022; Dhiman et al. 2022; Nekkanti et al. 2015). These formulations are considered liposome precursors because they can transform into liposomal dispersions upon hydration, enhancing drug stability, bioavailability, and therapeutic efficacy. Compared with traditional liposomes, proliposome formulations can offer improved stability and extended shelf life, minimizing the degradation risk during storage (Choudhary et al. 2022; Dhiman et al. 2022; Nekkanti et al. 2015). Additionally, proliposomes may generate liposomes that enable controlled and sustained drug release, making them highly appropriate for maintaining therapeutic levels over time (Choudhary et al. 2022; Dhiman et al. 2022; Nekkanti et al. 2015).

### 2.5.1 Advantages of Proliposomes Over the Thin-Film Hydration Method

Proliposome may transform into liposomes upon hydration, which simplifies storage and handling compared with the aqueous dispersions required in the thin-film hydration method (Choudhary et al. 2022; Dhiman et al. 2022; Nekkanti et al. 2015). This characteristic makes proliposomes particularly suitable for applications where stability and ease of transport are critical. In particular, the dry type of proliposomes allows for extended storage without significant changes in their properties (Choudhary et al. 2022; Dhiman et al. 2022; Nekkanti et al. 2015). By contrast, traditional liposomes prepared by thin film hydration are typically stored as aqueous solutions, making them highly prone to degradation over time (Lombardo and Kiselev 2022; Xiang and Cao 2018). Additionally, the dry nature of proliposomes reduces the risk of microbial contamination (Dhiman et al. 2022; Lombardo and Kiselev 2022; Xiang and Cao 2018). Furthermore, the free-flowing nature of proliposomes makes them easier to handle and transport than the liquid suspensions produced by the thin-film hydration method (Dhiman et al. 2022; Lombardo and Kiselev 2022; Xiang and Cao 2018). Thus, liposomes are generated from proliposomes only prior to administration through simple hydration (Choudhary et al. 2022; Dhiman et al. 2022; Nekkanti et al. 2015). Compared with liposomes prepared by thin-film hydration, proliposomes have also been shown to generate liposomes offering higher drug entrapment compared to liposomes prepared using traditional methods. For example, Khan et al.

described a novel approach for preparing proliposomes that offered several advantages over traditional liposomes prepared by the thin-film hydration method (Khan et al. 2015). Compared with liposomes prepared via the thin-film hydration, liposomes generated from particulate-based proliposomes achieved significantly greater drug entrapment efficiency (47.05% vs. 25.66%, respectively) (Khan et al. 2015).

### 2.5.2   Proliposome Technologies Feasibility Over the Freeze-Drying Method

Proliposomes may offer an economically feasible alternative to the freeze-drying method for liposome preparation because of their cost-effectiveness, simplicity, and efficiency in production and storage. The proliposome approach involves creating dry, free-flowing powder/granules that can be easily hydrated to form liposomes, which reduces the need for complex and energy-intensive processes associated with freeze-drying. The cost-efficiency of the proliposome technologies over the traditional freeze-drying approach for liposome formulation serves as a promising method with cost benefits and resource efficiencies. They offer significant advantages in terms of lower production and storage costs, simplified processing, and enhanced product stability (Adali et al. 2020; Justo and Moraes 2010; Kasper and Friess 2011; Maja et al. 2020; Singodia et al. 2012; Trucillo et al. 2020). By contrast, freeze-drying requires specialized equipment, prolonged processing times, and substantial energy inputs to ensure appropriate sublimation of water from frozen liposomal preparations (Adali et al. 2020; Kasper and Friess 2011). Additionally, proliposome manufacturing can streamline the production process by reducing the number of handling steps and the need for cryoprotectants, which are necessary for freeze-drying to prevent structural damage to liposomes (Adali et al. 2020; Kasper and Friess 2011).

### 2.6   Particulate-Based Proliposomes

### 2.6.1   Small-Scale Production

Particulate-based proliposomes, are the first proliposome system introduced in literature, and if proliposome type is not mentioned, the authors in this work and most others, mean the particulate-based type. These proliposomes consist of soluble carrier particles, such as carbohydrates, coated with phospholipids. When an aqueous phase is added, MLVs are spontaneously formed even if minimal or no shaking is provided (i.e., under static conditions) (Elhissi et al. 2015; Payne et al. 1986). Particulate-based proliposomes, as a means of generating liposomes, are effective for improving the therapeutic potential of drugs with poor bioavailability and for the pulmonary drug delivery applications (Elhissi et al. 2012; Khan et al. 2020, 2023; Omer et al. 2018).

During formulation, the selection of suitable carriers is essential, as it directly influences the stability and delivery efficiency of the proliposomes. For example, it has been reported that the use of various dispersion media and carrier types in spray-dried formulations affected the particle size distribution of the resultant liposomes, crystallinity of the formulation, and entrapment efficiency of the drug in the vesicles (Khan et al. 2023). In addition to the impact of carrier type, the incorporation of cholesterol in the formulation have been shown to generate liposomes capable of encapsulating the antifungal drug amphotericin B, with evidence of formulation ability to reduce toxicity and enhances tissue distribution of the drug, highlighting the advantage of the method of preparation and cholesterol inclusion (Singodia et al. 2012).

The rotary evaporation technique, illustrated in Fig. 2.1, is one of the most utilized methods for producing particulate-based proliposomes on a laboratory scale, particularly for generating liposomes with controlled lipid coatings (Elhissi and Taylor 2005; Shah et al. 2006; Xu et al. 2024). On a small scale, proliposomes can be prepared using a modified rotary evaporator with a feed-line tube, by placing the carrier particles in a round-bottom flask attached to the evaporator (Elhissi et al. 2015). A chloroform solution of the lipid constituents is injected dropwise onto the carrier particles through the feed line under reduced pressure (Elhissi et al. 2015). The solvent evaporates, creating proliposome granules that can either be immediately hydrated by adding an aqueous phase, or stored in a freezer for later use when needed (Elhissi et al. 2015; Payne et al. 1986). In this process within the rotary evaporator, the carrier particles such as sucrose, lactose, mannitol, sodium chloride, fructose, or glucose are coated with a lipid layer through dripping/spraying an organic solution over the carrier particles, followed by solvent evaporation under negative pressure (Elhissi et al. 2015; Elhissi and Taylor 2005; Shah et al. 2006; Xu et al. 2024).

## 2.6.2  Scaling Up Production of Particulate-Based Proliposomes

Scaling the production of proliposomes from the laboratory to the industrial scale involves adapting existing methods to produce larger quantities while maintaining the consistency, efficiency, and quality of the final product. This transition from small-scale techniques such as rotary evaporation to methods more suited for industrial-scale production, including fluidized-bed coating and spray draying, is crucial for meeting commercial demands in the pharmaceutical industry.

### 2.6.2.1  Fluidized-Bed Coating Method

Fluidized-bed coating is particularly effective for scaling up proliposome production because of its ability to process large batches of carrier particles with a uniform lipid coating. In this method, carrier particles such as sucrose or mannitol are fluidized in a chamber while a lipid solution is sprayed onto them, forming a lipid coat on the carrier

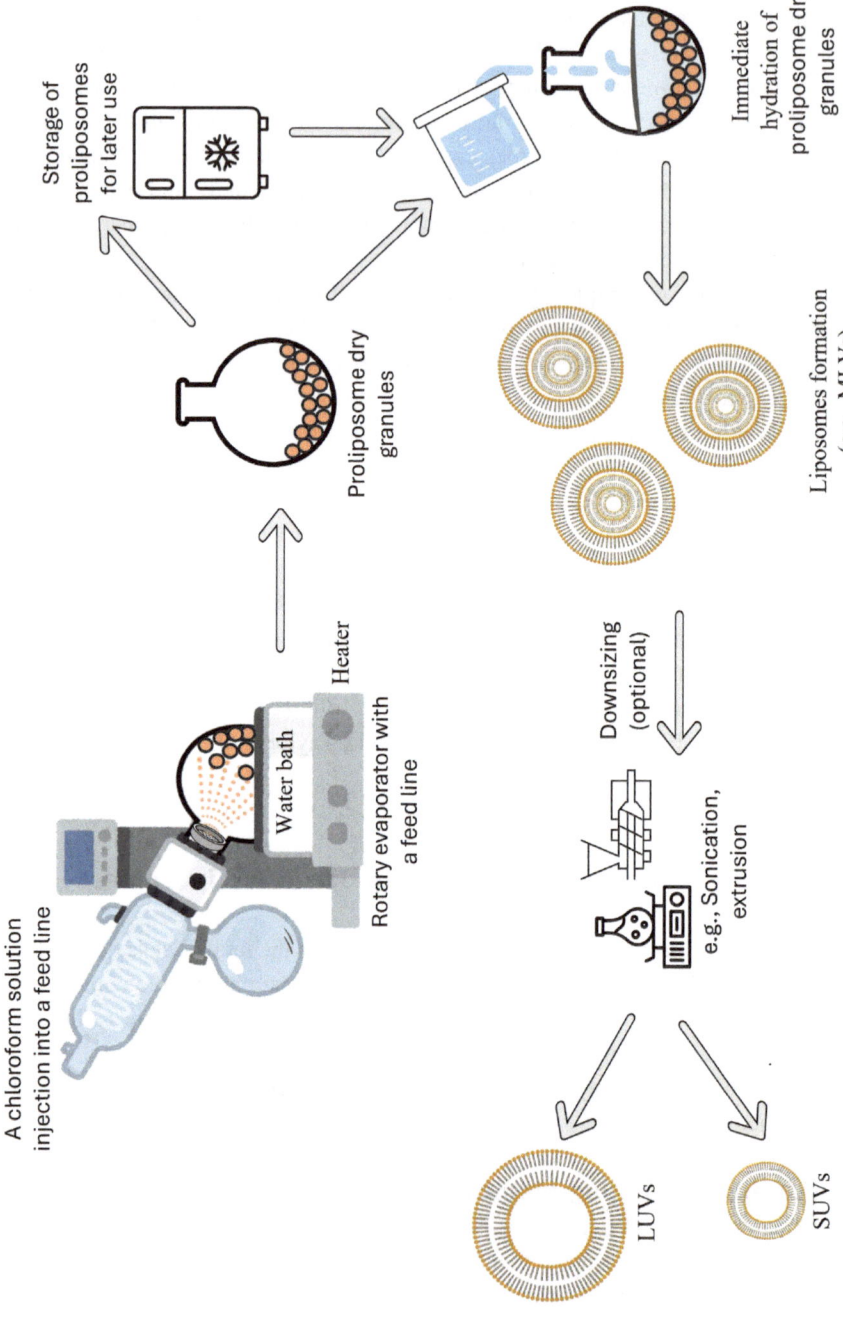

**Fig. 2.1** A schematic representation of the key steps involved in the small-scale production of particulate-based proliposomes using the modified rotary evaporation technique

particles as the solvent evaporates during fluidization (Chen and Alli 1987; Elhissi et al. 2015; Gala et al. 2015). This method is conducted with care since the high vapour pressure of organic solvents may cause environmental health (e.g. breathing toxic or irritant vapours) and safety issues. Thus, gradual and moderate quantities of ethanol were used (instead of chloroform that is commonly used with rotary evaporation) to allow safe deposition of the lipids onto the carbohydrate carrier particles. The process parameters, such as air pressure, spray rate, and temperature, must be precisely controlled, allowing for high reproducibility across batches (Chen and Alli 1987; Elhissi et al. 2015; Gala et al. 2015). Compared with traditional small-scale techniques, fluidized-bed coating is more efficient in handling larger volumes, reducing the time and labour involved in proliposome production. Unlike rotary evaporation, fluidized-bed coating is more efficient for large-scale production with consistent batch production and minimal manual intervention (Chen and Alli 1987; Elhissi et al. 2015; Gala et al. 2015). This method has proven effective for generating proliposomes intended for pulmonary and oral applications, as it allows for high encapsulation efficiencies and well-defined particle sizes (Dhiman et al. 2022; Gala et al. 2015; Nekkanti et al. 2015).

In our research group, the Fluidized-Bed Coating Method was utilized to produce proliposomes by coating sucrose particles with hydrogenated soya phosphatidylcholine and beclometasone dipropionate as a model antiasthma steroid (Gala et al. 2015). This method, illustrated in Fig. 2.2, aimed to enhance large-scale production of nano-liposome powders by employing an industrially viable approach (Gala et al. 2015). After coating, the proliposomes were hydrated, followed by size-reduction using high-pressure homogenization, and then freeze-drying (Gala et al. 2015). Importantly, freeze-drying was applied using a self-cryoprotected formulation since sucrose was used as a carrier for the manufacture of proliposomes. Findings demonstrated that high-pressure homogenization led to smaller liposome sizes compared to probe-sonication, yielding sizes between 69.84 and 89.34 nm (Gala et al. 2015). Freeze-drying and rehydration slightly increased liposome size but maintained it under 155 nm (Gala et al. 2015). Sucrose acted as both a carrier and a cryoprotectant, aiding in maintaining liposome stability and drug entrapment efficiency (Gala et al. 2015). This method offers a scalable, efficient approach for nano-liposome production using accessible industrial technology (Gala et al. 2015). The detailed process is illustrated in Fig. 2.2 (Gala et al. 2015).

### 2.6.2.2  Spray Drying

Spray drying has been used in the preparation of particulate-based proliposome because of its scalability, efficiency, and ability to produce dry, stable powders with excellent aerodynamic properties (Adel et al. 2021; Elhissi et al. 2015; Khan et al. 2023; Shreya et al. 2021). It involves spraying a mixture of drugs and lipids in a solvent through a spray dryer, which atomizes the solution into fine droplets and rapidly dries them to form low-density proliposome powders (Adel et al. 2021; Elhissi et al. 2015; Khan et al. 2023; Shreya et al. 2021).

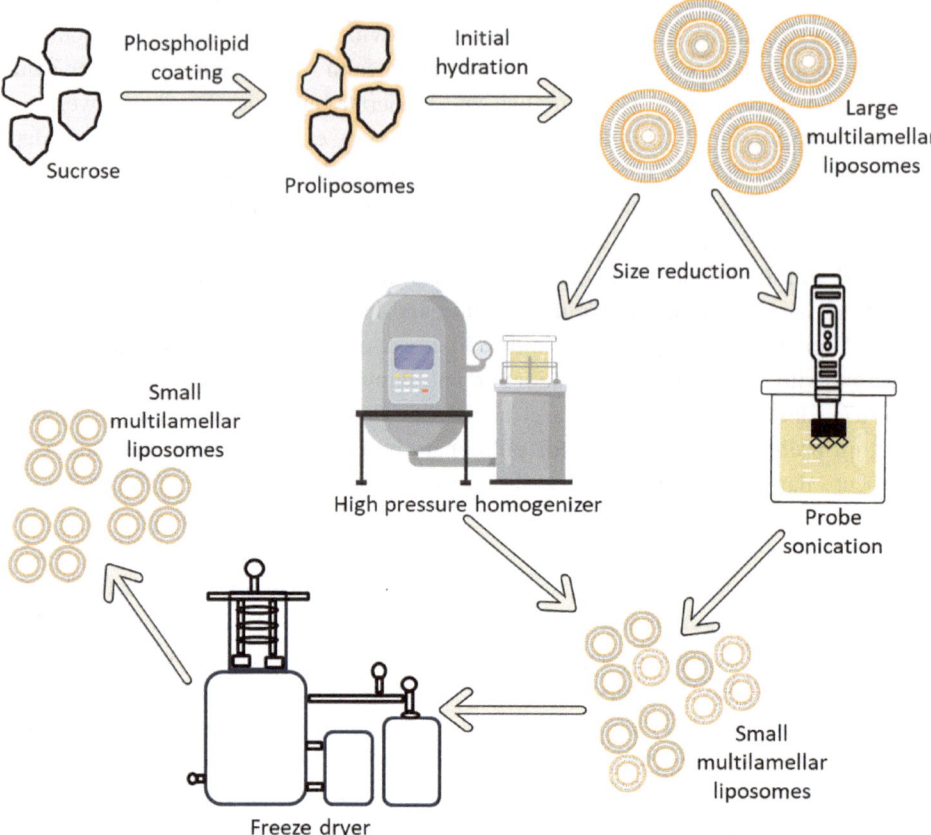

**Fig. 2.2** The large-scale production method employed by Gala et al. (2015) for the production of particulate-based proliposomes. Fluid-bed coating, high-pressure homogenization, and freeze-drying were performed

As depicted in Fig. 2.3, the process begins with the preparation of a lipid mixture, typically comprising phospholipids such as soya phosphatidylcholine and cholesterol, dissolved in an organic solvent such as ethanol (Adel et al. 2021; Elhissi et al. 2015; Khan et al. 2023; Shreya et al. 2021). The active ingredient is incorporated into this lipid solution, often using a surfactant or other excipient to enhance dispersion (Adel et al. 2021; Elhissi et al. 2015; Khan et al. 2023; Shreya et al. 2021). The lipid mixture is then sprayed through a nozzle, which atomizes it into fine droplets (Adel et al. 2021; Elhissi et al. 2015; Khan et al. 2023; Shreya et al. 2021). Spray drying parameters, such as inlet and outlet temperatures, airflow rates, and nozzle size, are critical parameters for determining particle size, encapsulation efficiency, and production yield of the resultant powder. As the solvent evaporates, the lipid and drug form a proliposome powder (Adel et al. 2021; Elhissi et al. 2015; Khan et al. 2023; Shreya et al. 2021). These proliposomes are more stable than liquid liposome formulations because they are in dry powder form, which reduces the risk of lipid oxidation and hydrolysis (Adel et al. 2021; Elhissi et al. 2015; Khan et al. 2023; Shreya et al. 2021).

Omer et al. prepared proliposomes using spray drying for possible aerosol delivery to the lungs using a Büchi B-290 spray drier (Omer et al. 2018). Spray-dried mannitol or lactose monohydrate (LMH) microparticles served as core carriers in the proliposome production process (Omer et al. 2018). This involved measuring a total of 100 mg lipid composed of soy phosphatidylcholine (SPC) and cholesterol (CH) (1:1 mol/mol), then adding 100 mL of 96% ethanol and 10 mg of salbutamol sulphate (Omer et al. 2018). The mixture was sonicated for 1 min to ensure complete lipid dissolution in ethanol

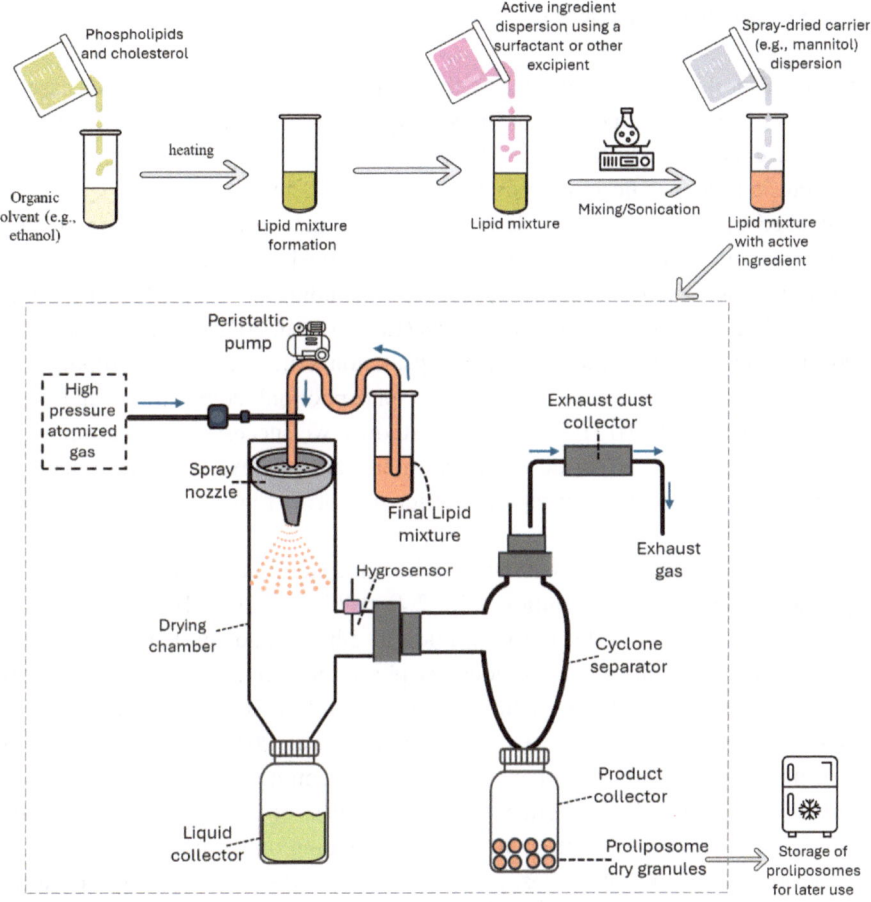

**Fig. 2.3** The common steps involved in the production of particulate-based proliposomes using the spray drying method

(Omer et al. 2018). Various ratios of spray-dried mannitol or LMH were then dispersed in the ethanolic solution and sonicated for 15 min to disintegrate any carbohydrate particle clumps before undergoing spray drying with the Büchi B-290 Mini Spray Dryer, now connected to Büchi's inert loop system (Omer et al. 2018). Continuous stirring maintained the uniformity of the alcoholic mixture as it was fed into the spray dryer (Omer et al. 2018). Optimal spray drying conditions were achieved by setting the inlet temperature to 120 °C, spray flow rate to 600 L/h, feed rate to 11%, and outlet temperature to $73 \pm 3$ °C (Omer et al. 2018). The resulting powder represented the final product (proliposome powder), which was collected and stored in a desiccator for later use (Omer et al. 2018).

## 2.7    Solvent-Based Proliposomes

### 2.7.1    Formulation Process

Solvent-based proliposomes, particularly those utilizing ethanol (i.e. ethanol-based proliposomes), is a convenient method for liposome preparation. These proliposomes typically consist of a concentrated ethanolic solution containing phospholipids, and possibly an aqueous phase at a specific low ratio (5:4:10 w/w/w) (Elhissi et al. 2015). This mixture forms stacked bilayers or simple ethanolic solutions that readily convert into liposomes upon dilution with an aqueous medium (Perrett et al. 1991; Elhissi et al. 2015). The resulting liposomes can exhibit various morphologies, including OLVs, MLVs, or a mixture of both, depending on the preparation conditions (Elhissi et al. 2015). Notably, ethanol-based proliposomes have demonstrated high encapsulation efficiency for hydrophilic materials, reaching up to 80%, depending on the phospholipid composition and hydration procedure (Elhissi et al. 2015). In one of our studies, we successfully encapsulated salbutamol sulphate in liposomes made from equimolar ratio of soya phosphatidylcholine and cholesterol, resulting in drug entrapment efficiency of as high as 62.07% (Elhissi et al. 2006).

A phospholipid such as SPC without or with cholesterol (1:1) are dissolved in ethanol with gentle heating to aid dissolution (Fig. 2.4) (Elhissi et al. 2010, 2011; Najlah et al. 2018; Perrett et al. 1991). The ratio of lipids to ethanol and the temperature used can vary depending on formulation. An aqueous phase, containing a water-soluble drug, is then added to the ethanolic lipid solution, usually while it is still hot enough to prevent lipid solidification (Elhissi et al. 2010, 2011; Najlah et al. 2018; Perrett et al. 1991). The resulting mixture is vigorously mixed to form a concentrated lipid formulation, which is then diluted with an additional aqueous phase (Elhissi et al. 2010, 2011; Najlah et al. 2018; Nguyen et al. 2024; Perrett et al. 1991). Depending on the intended application and desired properties of the liposomes, an optional size reduction step may be performed. Probe-sonication or extrusion through polycarbonate membranes can be used to achieve the desired size distribution.

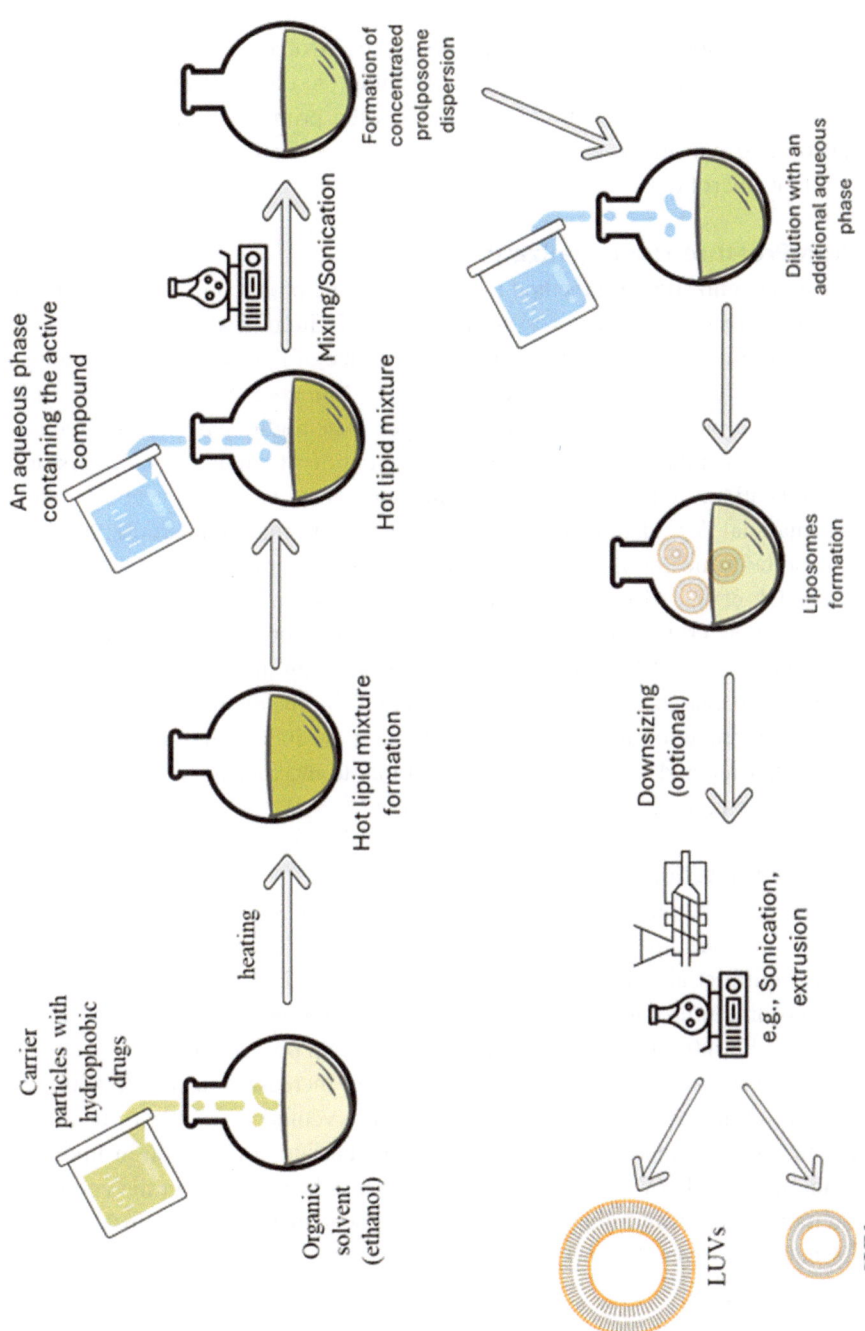

**Fig. 2.4**   Schematic representation of the key steps involved in the small-scale production of solvent-based proliposomes

## 2.7.2    Scaling Up Using the Solvent-Based Proliposome Method

The large-scale preparation of solvent-based proliposomes involves several advanced techniques that ensure efficient production while maintaining the quality and stability of the liposomes. These methods are mainly used to process the primary formulations of proliposome/liposome to uniformly downsize the vesicles at a relatively larger scale using homogenization or microfluidization.

### 2.7.2.1    High-Pressure Homogenization

High-pressure homogenization is a method used for liposome production, where lipids are subjected to intense mechanical forces to form vesicles (Beltrán et al. 2020). This technique, illustrated in Fig. 2.5, is effective in reducing the size of liposomes and achieving a uniform/narrow size distribution (Hızır-Kadı et al. 2020; Najlah et al. 2015, 2019). High-pressure homogenization offers significant advantages over traditional methods such as probe sonication and membrane extrusion, for the preparation of nanoparticle dispersions. Unlike probe sonication, high-pressure homogenization avoids excessive heating, reducing the risk of material degradation and contamination (due to titanium particles leaching from titanium probe-sonicators), which is critical for formulations intended for in vivo applications (Najlah et al. 2015). Additionally, it overcomes the limitations of the extrusion technique, such as prolonged processing times and difficulty in reducing the size of rigid vesicles (Najlah et al. 2015). The method involves passing the liposomal suspension through an interaction chamber at high pressure, where the collision of two streams of the suspension causes a reduction in vesicle size (Beltrán et al. 2020). Homogenization can be followed by other methods, such as microfluidization, to increase the physical stability of liposomes (Beltrán et al. 2020).

### 2.7.2.2    Microfluidization

Microfluidization is a homogenization technique that processes emulsions through a Microfluidizer to produce liposomes with small and uniform sizes (Fig. 2.6) (Kosaraju et al. 2006; McAuliffe et al. 2016; Nongonierma et al. 2009, 2013; Vemuri et al. 1990). Unlike other liposome preparation techniques, microfluidization eliminates the need for organic solvents and is advantageous for its ability to scale up production from the laboratory to industrial levels, as demonstrated by the successful increase from 100 to 4000 mL batches (Larivière et al. 1991; Nongonierma et al. 2013; Vemuri et al. 1990). It utilizes high pressure to force the fluid through microchannels, causing a reduction in liposome size to the nanometre scale (Vemuri et al. 1990). This is achieved through the combined effects of cavitation, shear, and impact within the microchannels, resulting in a uniform dispersion of nanoliposomes (Vemuri et al. 1990).

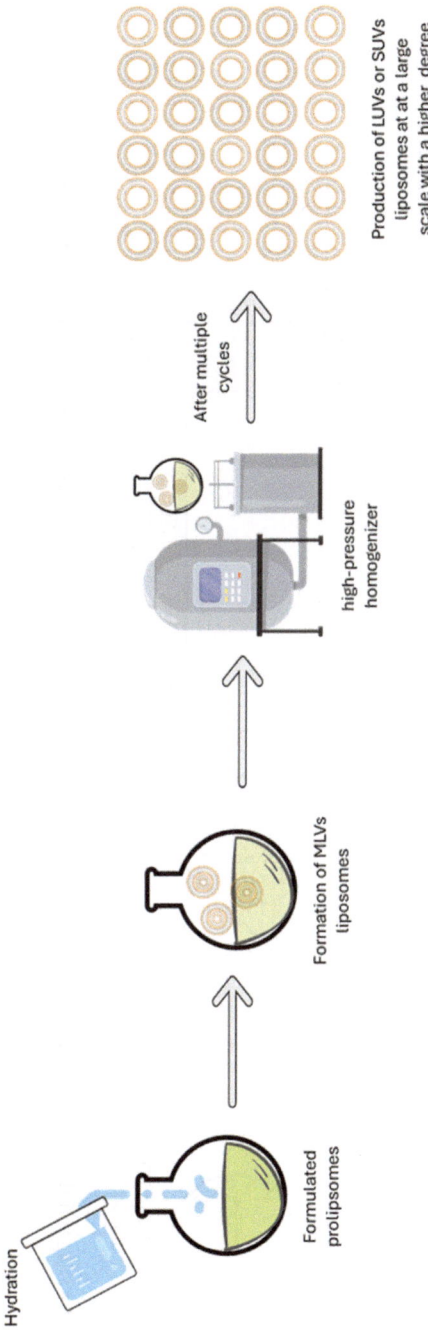

**Fig. 2.5**  The common steps involved in the large-scale production of liposomes from proliposomes using the high-pressure homogenization process

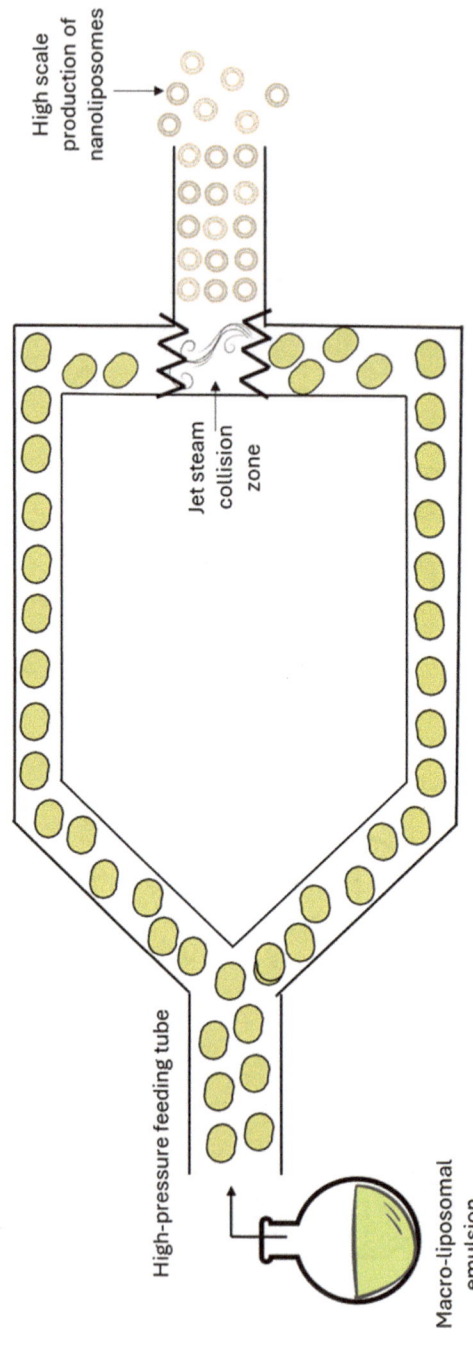

High scale
production of
nanoliposomes

Jet steam
collision
zone

High-pressure feeding tube

Macro-liposomal
emulsion

**Fig. 2.6** Diagram illustrating the common steps involved in the large-scale production of liposomes from proliposomes using the microfluidization process

## 2.8 Conclusion

Proliposomes represent a highly promising drug delivery system with several advantages over conventional liposomes. These advantages include improved stability, as proliposomes are more resistant to degradation and can be stored as a dry powder, which reduces the risks of hydrolysis, oxidation and contamination. Proliposome technologies allow for easier handling and transport and long-term storage. Moreover, the ability of proliposomes to generate liposomes upon hydration provides a useful delivery system adaptable to a range of therapeutic needs. This adaptability enhances the potential for proliposomes to be tailored to various drugs and diseases. Key factors in the optimization process, such as the choice of suitable carrier particles, solvents, lipid composition, and process parameters, are critical for achieving high drug encapsulation efficiency, controlled release, and consistent liposome particle size. Large-scale manufacture of particulate-based proliposomes can be achieved using fluidized-bed coating, and large-scale manufacture of liposomes generated from proliposomes can be achieved using methods like high-pressure homogenization and microfluidization.

## References

Adali, M. B., Barresi, A. A., Boccardo, G., & Pisano, R. (2020). Spray Freeze-Drying as a Solution to Continuous Manufacturing of Pharmaceutical Products in Bulk. *Processes*, *8*(6), 709. https://doi.org/10.3390/pr8060709

Adel, I. M., Elmeligy, M. F., Abdelrahim, M. E. A., Maged, A., Abdelkhalek, A. A., Abdelmoteleb, A. M. M., & Elkasabgy, N. A. (2021). Design and characterization of spray-dried proliposomes for the pulmonary delivery of curcumin. *International Journal of Nanomedicine*, *16*, 2667–2687. https://doi.org/10.2147/IJN.S306831

Akbarzadeh, A., Rezaei-Sadabady, R., Davaran, S., Joo, S. W., Zarghami, N., Hanifehpour, Y., Samiei, M., Kouhi, M., & Nejati-Koshki, K. (2013). Liposome: classification, preparation, and applications. *Nanoscale Research Letters*, *8*(1), 102. https://doi.org/10.1186/1556-276X-8-102

Andra, V. V. S. N. L., Pammi, S. V. N., Bhatraju, L. V. K. P., & Ruddaraju, L. K. (2022). A Comprehensive Review on Novel Liposomal Methodologies, Commercial Formulations, Clinical Trials and Patents. *BioNanoScience*, *12*(1), 274–291. https://doi.org/10.1007/s12668-022-00941-x

Ball, R., Bajaj, P., & Whitehead, K. (2016). Achieving long-term stability of lipid nanoparticles: examining the effect of pH, temperature, and lyophilization. *International Journal of Nanomedicine*, *12*, 305–315. https://doi.org/10.2147/IJN.S123062

Beltrán, J. D., Ricaurte, L., Estrada, K. B., & Quintanilla-Carvajal, M. X. (2020). Effect of homogenization methods on the physical stability of nutrition grade nanoliposomes used for encapsulating high oleic palm oil. *LWT*, *118*. https://doi.org/10.1016/j.lwt.2019.108801

Boafo, G. F., Magar, K. T., Ekpo, M. D., Qian, W., Tan, S., & Chen, C. (2022). The Role of Cryoprotective Agents in Liposome Stabilization and Preservation. In *International Journal of Molecular Sciences* (Vol. 23, Issue 20). MDPI. https://doi.org/10.3390/ijms232012487

Boban, Z., Mardešić, I., Subczynski, W. K., & Raguz, M. (2021). Giant Unilamellar Vesicle Electroformation: What to Use, What to Avoid, and How to Quantify the Results. *Membranes*, *11*(11), 860. https://doi.org/10.3390/membranes11110860

Chaurasiya, A., Gorajiya, A., Panchal, K., Katke, S., & Singh, A. K. (2022). A review on multivesic-ular liposomes for pharmaceutical applications: preparation, characterization, and translational challenges. *Drug Delivery and Translational Research*, *12*(7), 1569–1587. https://doi.org/10.1007/s13346-021-01060-y

Chen, C.-M., & Alli, D. (1987). Use of Fluidized Bed in Proliposome Manufacturing. *Journal of Pharmaceutical Sciences*, *76*(5), 419. https://doi.org/10.1002/jps.2600760517

Choudhary, M., Chaurawal, N., Barkat, Md. A., & Raza, K. (2022). Proliposome-Based Nanostrate-gies: Challenges and Development as Drug Delivery Systems. *AAPS PharmSciTech*, *23*(8), 293. https://doi.org/10.1208/s12249-022-02443-1

de Freitas, C. F., Calori, I. R., Tessaro, A. L., Caetano, W., & Hioka, N. (2019). Rapid formation of Small Unilamellar Vesicles (SUV) through low-frequency sonication: An innovative approach. *Colloids and Surfaces B: Biointerfaces*, *181*, 837–844. https://doi.org/10.1016/j.colsurfb.2019.06.027

Dhiman, N., Sarvaiya, J., & Mohindroo, P. (2022). A drift on liposomes to proliposomes: recent advances and promising approaches. *Journal of Liposome Research*, *32*(4), 317–331. https://doi.org/10.1080/08982104.2021.2019762

Elhissi, A. (2017). Liposomes for Pulmonary Drug Delivery: The Role of Formulation and Inhalation Device Design. *Current Pharmaceutical Design*, *23*(3), 362–372. https://doi.org/10.2174/1381612823666161116114732

Elhissi, A. M. A., & Taylor, K. M. G. (2005). Delivery of liposomes generated from proliposomes using air-jet, ultrasonic, and vibrating-mesh nebulisers. *Journal of Drug Delivery Science and Technology*, *15*(4), 261–265. https://doi.org/10.1016/S1773-2247(05)50047-9

Elhissi, A., Karnam, K., Danesh-Azari, M.-R., Gill, H., & Taylor, K. (2006). Formulations generated from ethanol-based proliposomes for delivery via medical nebulizers. *Journal of Pharmacy and Pharmacology*, *58*(7), 887–894. https://doi.org/10.1211/jpp.58.7.0002

Elhissi, A. M. A., Karnam, K. K., Danesh-Azari, M.-R., Gill, H. S., & Taylor, K. M. G. (2010). Formulations generated from ethanol-based proliposomes for delivery via medical nebulizers. *Journal of Pharmacy and Pharmacology*, *58*(7), 887–894. https://doi.org/10.1211/jpp.58.7.0002

Elhissi, A., Gill, H., Ahmed, W., & Taylor, K. (2011). Vibrating-mesh nebulization of liposomes generated using an ethanol-based proliposome technology. *Journal of Liposome Research*, *21*(2), 173–180. https://doi.org/10.3109/08982104.2010.505574

Elhissi, A. M. A., Ahmed, W., & Taylor, K. M. G. (2012). Laser diffraction and electron microscopy studies on inhalable liposomes generated from particulate-based proliposomes within a medical Nebulizer. *Journal of Nanoscience and Nanotechnology*, *12*(8), 6693–6699. https://doi.org/10.1166/jnn.2012.4566

Elhissi, A., Phoenix, D., & Ahmed, W. (2015). Chapter 15 - Some approaches to large-scale man-ufacturing of liposomes. In W. Ahmed & M. J. Jackson (Eds.), *Emerging Nanotechnologies for Manufacturing (Second Edition)* (pp. 402–417). William Andrew Publishing. https://doi.org/10.1016/B978-0-323-28990-0.00015-4

Gala, R. P., Khan, I., Elhissi, A. M. A., & Alhnan, M. A. (2015). A comprehensive production method of self-cryoprotected nano-liposome powders. *International Journal of Pharmaceutics*, *486*(1–2), 153–158. https://doi.org/10.1016/j.ijpharm.2015.03.038

Ghanbarzadeh, S., Valizadeh, H., & Zakeri-Milani, P. (2013). The effects of lyophilization on the physico-chemical stability of sirolimus liposomes. *Advanced Pharmaceutical Bulletin*, *3*(1), 25–29. https://doi.org/10.5681/apb.2013.005

Giuliano, C. B., Cvjetan, N., Ayache, J., & Walde, P. (2021). Multivesicular Vesicles: Preparation and Applications. *ChemSystemsChem*, *3*(2). https://doi.org/10.1002/syst.202000049

Hızır-Kadı, İ., Gültekin-Özgüven, M., Altin, G., Demircan, E., & Özçelik, B. (2020). Liposomal nanodelivery systems generated from proliposomes for pollen extract with improved solubility and in vitro bioaccessibility. *Heliyon*, *6*(9), e05030. https://doi.org/10.1016/j.heliyon.2020.e05030

Justo, O. R., & Moraes, Â. M. (2010). Economical Feasibility Evaluation of an Ethanol Injection Liposome Production Plant. *Chemical Engineering & Technology*, *33*(1), 15–20. https://doi.org/10.1002/ceat.200800502

Kasper, J. C., & Friess, W. (2011). The freezing step in lyophilization: Physico-chemical fundamentals, freezing methods and consequences on process performance and quality attributes of biopharmaceuticals. *European Journal of Pharmaceutics and Biopharmaceutics*, *78*(2), 248–263. https://doi.org/10.1016/j.ejpb.2011.03.010

Khan, I., Yousaf, S., Subramanian, S., Korale, O., Alhnan, M. A., Ahmed, W., Taylor, K. M. G., & Elhissi, A. (2015). Proliposome powders prepared using a slurry method for the generation of beclometasone dipropionate liposomes. *International Journal of Pharmaceutics*, *496*(2), 342–350. https://doi.org/10.1016/j.ijpharm.2015.10.002

Khan, I., Lau, K., Bnyan, R., Houacine, C., Roberts, M., Isreb, A., Elhissi, A., & Yousaf, S. (2020). A Facile and Novel Approach to Manufacture Paclitaxel-Loaded Proliposome Tablet Formulations of Micro or Nano Vesicles for Nebulization. *Pharmaceutical Research*, *37*(6). https://doi.org/10.1007/s11095-020-02840-w

Khan, I., Al-Hasani, A., Khan, M. H., Khan, A. N., Fakhr-E-Alam, Sadozai, S. K., Elhissi, A., Khan, J., & Yousaf, S. (2023). Impact of dispersion media and carrier type on spray-dried proliposome powder formulations loaded with beclomethasone dipropionate for their pulmonary drug delivery via a next generation impactor. *PLoS ONE*, *18*(3 March). https://doi.org/10.1371/journal.pone.0281860

Kosaraju, S. L., Tran, C., & Lawrence, A. (2006). Liposomal Delivery Systems for Encapsulation of Ferrous Sulfate: Preparation and Characterization. *Journal of Liposome Research*, *16*(4), 347–358. https://doi.org/10.1080/08982100600992351

Koudelka, Š., Mašek, J., Neuzil, J., & Turánek, J. (2010). Lyophilised liposome-based formulations of α-tocopheryl succinate: Preparation and physico-chemical characterisation. *Journal of Pharmaceutical Sciences*, *99*(5), 2434–2443. https://doi.org/10.1002/jps.22002

Larivière, B., El Soda, M., Soucy, Y., Trépanier, G., Paquin, P., & Vuillemard, J. C. (1991). Microfluidized liposomes for the acceleration of cheese ripening. *International Dairy Journal*, *1*(2), 111–124. https://doi.org/10.1016/0958-6946(91)90003-Q

Liu, P., Chen, G., & Zhang, J. (2022). A Review of Liposomes as a Drug Delivery System: Current Status of Approved Products, Regulatory Environments, and Future Perspectives. *Molecules*, *27*(4), 1372. https://doi.org/10.3390/molecules27041372

Lombardo, D., & Kiselev, M. A. (2022). Methods of Liposomes Preparation: Formation and Control Factors of Versatile Nanocarriers for Biomedical and Nanomedicine Application. *Pharmaceutics*, *14*(3), 543. https://doi.org/10.3390/pharmaceutics14030543

Lu, W.-L., & Qi, X.-R. (Eds.). (2021). *Liposome-Based Drug Delivery Systems*. Springer Berlin Heidelberg. https://doi.org/10.1007/978-3-662-49320-5

Maja, L., Željko, K., & Mateja, P. (2020). Sustainable technologies for liposome preparation. *The Journal of Supercritical Fluids*, *165*, 104984. https://doi.org/10.1016/j.supflu.2020.104984

McAuliffe, L. N., Kilcawley, K. N., Sheehan, J. J., & McSweeney, P. L. H. (2016). Manufacture and Incorporation of Liposome-Entrapped Ethylenediaminetetraacetic Acid into Model Miniature Gouda-Type Cheese and Subsequent Effect on Starter Viability, pH, and Moisture Content. *Journal of Food Science*, *81*(11). https://doi.org/10.1111/1750-3841.13519

Naeem, A. (2017). Liposomes: A Novel Drug Delivery System. *SSRN Electronic Journal*. https://doi.org/10.2139/ssrn.2960975

Nair, K. S., & Bajaj, H. (2023). Advances in giant unilamellar vesicle preparation techniques and applications. *Advances in Colloid and Interface Science, 318*, 102935. https://doi.org/10.1016/j.cis.2023.102935

Najlah, M., Hidayat, K., Omer, H. K., Mwesigwa, E., Ahmed, W., AlObaidy, K. G., Phoenix, D. A., & Elhissi, A. (2015). A facile approach to manufacturing non-ionic surfactant nanodipsersions using proniosome technology and high-pressure homogenization. *Journal of Liposome Research, 25*(1), 32–37. https://doi.org/10.3109/08982104.2014.924140

Najlah, M., Jain, M., Wan, K. W., Ahmed, W., Albed Alhnan, M., Phoenix, D. A., Taylor, K. M. G., & Elhissi, A. (2018). Ethanol-based proliposome delivery systems of paclitaxel for in vitro application against brain cancer cells. *Journal of Liposome Research, 28*(1), 74–85. https://doi.org/10.1080/08982104.2016.1259628

Najlah, M., Said Suliman, A., Tolaymat, I., Kurusamy, S., Kannappan, V., Elhissi, A. M. A., & Wang, W. (2019). Development of Injectable PEGylated Liposome Encapsulating Disulfiram for Colorectal Cancer Treatment. *Pharmaceutics, 11*(11), 610. https://doi.org/10.3390/pharmaceutics11110610

Nekkanti, V., Venkatesan, N., & Betageri, G. (2015). Proliposomes for Oral Delivery: Progress and Challenges. *Current Pharmaceutical Biotechnology, 16*(4), 303–312. https://doi.org/10.2174/1389201016666150118134256

Neves, P., Lopes, S. C. D. N., Sousa, I., Garcia, S., Eaton, P., & Gameiro, P. (2009). Characterization of membrane protein reconstitution in LUVs of different lipid composition by fluorescence anisotropy. *Journal of Pharmaceutical and Biomedical Analysis, 49*(2), 276–281. https://doi.org/10.1016/j.jpba.2008.11.026

Nguyen, D. T., Kim, M. H., Baek, M. J., Kang, N. W., & Kim, D. D. (2024). Preparation and evaluation of proliposomes formulation for enhancing the oral bioavailability of ginsenosides. *Journal of Ginseng Research, 48*(4), 417–424. https://doi.org/10.1016/j.jgr.2024.03.004

Nongonierma, A. B., Abrlova, M., Fenelon, M. A., & Kilcawley, K. N. (2009). Evaluation of Two Food Grade Proliposomes To Encapsulate an Extract of a Commercial Enzyme Preparation by Microfluidization. *Journal of Agricultural and Food Chemistry, 57*(8), 3291–3297. https://doi.org/10.1021/jf803367b

Nongonierma, A., Abrlova, M., & Kilcawley, K. (2013). Encapsulation of a Lactic Acid Bacteria Cell-Free Extract in Liposomes and Use in Cheddar Cheese Ripening. *Foods, 2*(1), 100–119. https://doi.org/10.3390/foods2010100

Nsairat, H., Khater, D., Sayed, U., Odeh, F., Al Bawab, A., & Alshaer, W. (2022). Liposomes: structure, composition, types, and clinical applications. *Heliyon, 8*(5), e09394. https://doi.org/10.1016/j.heliyon.2022.e09394

Omer, H. K., Hussein, N. R., Ferraz, A., Najlah, M., Ahmed, W., Taylor, K. M. G., & Elhissi, A. M. A. (2018). Spray-Dried Proliposome Microparticles for High-Performance Aerosol Delivery Using a Monodose Powder Inhaler. *AAPS PharmSciTech, 19*(5), 2434–2448. https://doi.org/10.1208/s12249-018-1058-4

Payne, N. I., Browning, I., & Hynes, C. A. (1986). Characterization of Proliposomes. *Journal of Pharmaceutical Sciences, 75*(4), 330–333. https://doi.org/10.1002/jps.2600750403

Perrett, S., Golding, M., & Williams, W. P. (1991). A Simple Method for the Preparation of Liposomes for Pharmaceutical Applications: Characterization of the Liposomes. *Journal of Pharmacy and Pharmacology, 43*(3), 154–161. https://doi.org/10.1111/j.2042-7158.1991.tb06657.x

Sainaga Jyothi, V. G. S., Bulusu, R., Venkata Krishna Rao, B., Pranothi, M., Banda, S., Kumar Bolla, P., & Kommineni, N. (2022). Stability characterization for pharmaceutical liposome product development with focus on regulatory considerations: An update. *International Journal of Pharmaceutics, 624*, 122022. https://doi.org/10.1016/j.ijpharm.2022.122022

Sercombe, L., Veerati, T., Moheimani, F., Wu, S. Y., Sood, A. K., & Hua, S. (2015). Advances and Challenges of Liposome Assisted Drug Delivery. *Frontiers in Pharmacology, 6.* https://doi.org/10.3389/fphar.2015.00286

Shah, N. M., Parikh, J., Namdeo, A., Subramanian, N., & Bhowmick, S. (2006). Preparation, Characterization and *In Vivo* Studies of Proliposomes Containing Cyclosporine A. *Journal of Nanoscience and Nanotechnology, 6*(9), 2967–2973. https://doi.org/10.1166/jnn.2006.403

Shreya, A. B., Pandey, A., Nikam, A. N., Patil, P. O., Sonawane, R., Deshmukh, P. K., & Mutalik, S. (2021). One-pot development of spray dried cationic proliposomal dry powder insufflation: Optimization, characterization and bio-interactions. *Journal of Drug Delivery Science and Technology, 61.* https://doi.org/10.1016/j.jddst.2020.102298

Singodia, D., Verma, A., Khare, P., Dube, A., Mitra, K., & Mishra, P. R. (2012). Investigations on feasibility of *in situ* development of amphotericin B liposomes for industrial applications. *Journal of Liposome Research, 22*(1), 8–17. https://doi.org/10.3109/08982104.2011.584317

Trucillo, P., Campardelli, R., Iuorio, S., De Stefanis, P., & Reverchon, E. (2020). Economic Analysis of a New Business for Liposome Manufacturing Using a High-Pressure System. *Processes, 8*(12), 1604. https://doi.org/10.3390/pr8121604

Vemuri, S., Yu, C.-D., Wangsatorntanakun, V., Roosdorp, N., Myer, B., & Westwood, N. (1990). *Large-Scale Production of Liposomes By a Microfluidizer.* https://doi.org/10.3109/03639049009043797

Xiang, B., & Cao, D.-Y. (2018). Preparation of Drug Liposomes by Thin-Film Hydration and Homogenization. In *Liposome-Based Drug Delivery Systems* (pp. 1–11). Springer Berlin Heidelberg. https://doi.org/10.1007/978-3-662-49231-4_2-1

Xu, L., Huang, Z., Pei, X., Zhang, Z., Li, S., & He, Y. (2024). Preparation and Characterization of Universal Liquid Proliposomes Encapsulating Water-soluble Efficacious Substances. *Current Pharmaceutical Biotechnology, 26.* https://doi.org/10.2174/0113892010328982241004064135

Zhang, G., & Sun, J. (2021). Lipid in Chips: A Brief Review of Liposomes Formation by Microfluidics. *International Journal of Nanomedicine, 16,* 7391–7416. https://doi.org/10.2147/IJN.S331639

Zhong, Q., & Zhang, H. (2023). *Preparation of Small Unilamellar Vesicle Liposomes Using Detergent Dialysis Method* (pp. 49–56). https://doi.org/10.1007/978-1-0716-2954-3_3

# Formulation Approaches for Proliposomes in Pulmonary Drug Delivery

**Abstract**

In this chapter, a range of characterization techniques for proliposomes and the resultant liposomes were elucidated, and their importance was discussed. For proliposomes, characterization techniques include X-ray diffraction to investigate powder crystallinity, particle morphology study using scanning electron microscopy (SEM) and angle of repose (AOR) to study proliposome flowability. These characteristics of proliposome powders are a result of the formulation design and can influence their aerosol performance, and the characteristics of the subsequently generated liposomes. Liposome characteristics are usually investigated in terms of particle size using laser diffraction (for micro-size vesicles) or dynamic light scattering (for nano-liposomes), zeta potential using electrophoretic mobility, and vesicle morphology and lamellarity using transmission electron microscopy (TEM). Proliposome formulations are delivered in inhalable aerosols either as proliposome powder, proliposome solution, or as liposome dispersion (following hydration of proliposomes). A limited number of studies were conducted on the delivery of proliposomes using pressurized metered dose inhalers (pMDIs), which was achieved by dissolving the lipid blend in chlorofluorocarbon (CFC) propellants in which the drug is dissolved or dispersed (depending on its propellant-solubility). Precise proliposome doses can be generated from pMDIs, resulting in immediate evaporation of the propellant and presumed hydration of lipids-drug blend into liposomes in situ by utilizing the aqueous environment of the lung. The replacement of the ozone-depleting CFC propellants with the non-ozone depleting hydrofluoroalkanes (HFAs) has limited the research of liposomes/proliposome delivery via pMDIs, because HFAs are poor solubilizes of phospholipids. Many research studies were conducted on the delivery of particulate-based proliposomes using dry powder inhalers (DPIs), mostly using spray drying to manufacture the proliposomes. Superiorly high proliposome aerosol performance could be achieved and assessed in vitro, depending on formulation composition. Medical nebulizers have shown a great capability of delivering liposomes generated from particulate-based or ethanol-based proliposomes. Proliposomes can utilize the shearing environment within nebulizer reservoir and convert in situ into inhalable liposomes. Air-jet and vibrating-mesh nebulizers have shown the ability to generate highly performing aerosols from this proliposome system, while the ultrasonic device delivered only 6% of the phospholipid content, suggesting its unsuitability for delivering this proliposome system. Many other studies of nebulizing liposomes generated from proliposome powders or dispersed tablets have been elucidated and evaluated in this chapter.

## 3.1    Introduction

Pulmonary drug delivery has emerged as a promising strategy for treating respiratory diseases and delivering systemic therapeutics with minimal side effects (Rudokas et al. 2016; Wang et al. 2024a). Proliposomes, as precursor systems capable of generating liposomes upon hydration, offer a useful and efficient method for pulmonary drug delivery (Elhissi 2017). Proliposome technologies may successfully address the challenges associated with traditional liposome formulations, owing to the relatively high stability of proliposomes, and the high encapsulation efficiency of the resultant liposomes upon hydration of proliposomes (Elhissi 2017).

As outlined in Chap. 2, various preparation techniques can be utilized for the production of particulate-based and solvent-based proliposomal formulations. The key components of proliposomal formulations include phospholipids, cholesterol, and water-soluble carriers (e.g. sucrose), along with other excipients, depending on the intended use of the formulation (Choudhary et al. 2022). To achieve a successful formulation design, careful selection of excipients and formulation strategies is essential, coupled with a thorough analytical evaluation and comprehensive characterization of the formulation.

This chapter highlights important characterization techniques for studying the physicochemical properties of proliposomal formulations and the generated liposomes. These investigations are considered necessary for ensuring the formulation efficacy, safety, stability and aerodynamic properties required for efficient pulmonary deposition of the inhaled formulation. Additionally, this chapter provides detailed insights into proliposome formulation approaches tailored for delivery from different types of inhalation devices, including formulations made for delivery via pressurized metered-dose inhalers (pMDIs), dry powder inhalers (DPIs), and medical nebulizers.

## 3.2 Proliposome Formulation Analysis and Characterization Techniques for Pulmonary Delivery

The physicochemical characteristics of a proliposomal product are influenced by factors such as structural design, preparation method, composition, and intended application (Khan et al. 2021; Lombardo and Kiselev 2022; Parhizkar et al. 2021). For proliposomes to function effectively as carriers of bioactive substances for pulmonary delivery, their formulation, including the resultant vesicle particle size, lamellarity, surface charge, drug encapsulation, and aerosolization performance, must be thoroughly assessed (Dallal Bashi et al. 2024; Elhissi et al. 2006a, b; Omer et al. 2018). Therefore, the evaluation of proliposomal formulations for pulmonary delivery requires comprehensive characterization techniques to determine their appropriateness and efficiency as drug carriers targeted to the lungs. Key characterization studies are discussed in details in the subsequent sections and summarized in Table 3.1.

### 3.2.1 Particle Size and Distribution

Particle size plays a crucial role in determining how proliposomes and liposomes behave when used for pulmonary drug delivery. The size of the particles affects their aerodynamic behavior and how they deposit in the respiratory tract (Elhissi 2017). Ideally, to be "therapeutically effective", aerosolized particles must have an aerodynamic size less than 5 μm in order to qualify for deposition in the "deep lung"; these particles can be described to be in the "fine particle fraction" (FPF) (Elhissi 2017). Particles smaller than 2 μm are particularly effective at reaching the alveolar region of the lung (Elhissi 2017). Dynamic light scattering (DLS), laser diffraction (LD), and conventional nanoparticle tracking analysis (NTA) are common techniques used to measure particle size (Singh et al. 2019). Newer approaches, such as multispectral advanced nanoparticle tracking analysis (MANTA), were also proposed as being more accurate than DLS, LD, and NTA in determining the particle size distribution in heterogeneous populations (Singh et al. 2019). The key common parameters of particle and aerosol size to be measured before or after pulmonary administration of proliposomes and liposomes are discussed below.

#### 3.2.1.1 Particle Size
##### Hydrodynamic Size (HDS)
The HDS is the intensity-weighted mean size of a particle as it diffuses through a liquid medium (Mišík et al. 2024). In an effort to understand how proliposomal DPI formulations may interact with alveolar macrophages and *Mycobacterium tuberculosis*, Srichana et al. formulated proliposome DPIs containing isoniazid and/or rifampicin and determined the hydrodynamic sizes following hydration (Srichana et al. 2022). Upon reconstitution of the DPIs, the HDS of the resultant nanoparticles ranged from 370.9 to 556.2 nm (Srichana

**Table 3.1** Summary of key analytical studies for the characterization of proliposomal formulations designed for pulmonary delivery

| Parameter | Analytical techniques | Key metrics |
|---|---|---|
| Particle size of liposomes | DLS, LD, NTA, MANTA | VMD, MMAD, Dae, span, GSD, PDI, HDS, FPF, FPD, RF |
| Zeta potential of liposomes | Zetasizer | ZP value |
| Surface morphology of proliposomes | SEM | Structural characteristics (e.g., uniformity, shape, patterns, defects, surface smoothness/roughness) |
| Morphology of liposomes | TEM, cryo-TEM | Shape of liposomes (spherical, elongated, etc.), lamellarity of liposomes (number of phospholipid bilayers), assessment of vesicle aggregation |
| Chemical structure of proliposome constituents (i.e., drug and excipients) | NMR, HPLC–MS, FTIR | Confirming chemical composition, molecular weight, and chemical bond interactions |
| Lipid concentration | Stewart assay | Lipid coating efficiency, phospholipid content, etc. |
| Diffraction patterns | XRD | Crystalline properties, phase transitions, melting points, enthalpy changes |
| Thermal behavior | DSC, high sensitivity DSC | Thermal profile of liposome vesicles, quality and quantity of drug and excipient interaction with the liposome bilayers, estimation of hydrophobic drug entrapment in liposomes |
| Drug entrapment, loading, and release | HPLC, HPLC–MS, UV–visible spectroscopy | %EE, LE, %ED |
| Proliposome flow properties | Carr's index | Bulk/tapped density, AOR |
| In vitro aerosol and nebulizer performance | Two-stage glass impinger, inhalation devices | Particle deposition and distribution, nebulization time, sputtering time, aerosol mass/phospholipid/drug output, aerosol mass/phospholipid/drug output rates |

(continued)

**Table 3.1**   (continued)

| Parameter | Analytical techniques | Key metrics |
|---|---|---|
| In vitro therapeutic efficacy and toxicity | MTT/XTT assays, disease specifics cell lines and assays | Cell viability, $IC_{50}$, and morphology, other disease specific outcomes |
| In vivo therapeutic efficacy and toxicity | Animal models | Pharmacokinetics, pharmacodynamics, adverse effects studies |

*%ED* percent emitted dose, *%EE* percent entrapment efficiency, *AOR* angle of repose, *Dae* aerodynamic diameter, *DLS* dynamic light scattering, *DSC* differential scanning calorimetry, *FPD* fine particle dose, *FPF* fine particle fraction, *FTIR* Fourier transform infrared spectroscopy, *GSD* geometric standard deviation, *HDS* hydrodynamic index, *HPLC–MS* high-performance liquid chromatography–mass spectrometry, *IC$_{50}$* half maximal inhibitory concentration, *LD* laser diffraction, *LD50* median lethal dose, *LE* loading efficiency, *MANTA* multispectral advanced nanoparticle tracking analysis, *MDI* metered-dose inhaler, *MMAD* mass median aerodynamic diameter, *MTT* 3-(4,5-dimethylthiazol-2-yl)-2,5-diphenyltetrazolium bromide, *NMR* nuclear magnetic resonance, *NTA* nanoparticle tracking analysis, *PDI* polydispersity index, *RF* respirable fraction, *SEM* scanning electron microscopy, *TEM* transmission electron microscopy, *VMD* volume median diameter, *XRD* X-ray diffraction, *XTT* 2,3-bis-(2-methoxy-4-nitro-5-sulfophenyl)-2H-tetrazolium-5-carboxanilide, *ZP* zeta potential

et al. 2022). In another study, Rojanarat et al. reported that levofloxacin proliposomes have generated liposomes with HDS in the submicron range, and demonstrated the ability to reduce the viability of *Mycobacterium bovis* residing in alveolar macrophages (Rojanarat et al. 2012b).

## Volume Median Diameter (VMD)

Volume median diameter (VMD) represents the particle diameter at which half of the total sample (by volume) is smaller and the other half is larger. In one study, salbutamol sulfate-loaded proliposomes were prepared via ethanol-based proliposome and hydrated for delivery via vibrating-mesh nebulization (Elhissi et al. 2013). The study compared the aerosol properties, including droplet size distribution, droplet VMD, aerosol output, and FPF of the aerosolized liposome formulation with those of a conventional drug solution (Elhissi et al. 2013). The results showed that the proliposome formulation outperformed the conventional solution in terms of output and eligibility for peripheral airways deposition, with FPF values of 57.85% and 45.81%, respectively (Elhissi et al. 2013).

## Aerodynamic Diameter (Dae)

Aerodynamic diameter (Dae) is another size metric that represents the diameter of a spherical particle with a standard reference density, typically 1 $g/cm^3$, that settles in air at the same vertical velocity as the particle in question (Finlay and Darquenne 2020). Particles with Dae > 5 $\mu$m would deposit in the upper respiratory tract, whereas particles with Dae 1–5 μm would settle in the bronchi and alveolar regions (Knap et al. 2023). Particles having Dae smaller than 5 μm are essential for deposition in the "deep lung" and hence would be considered "therapeutically useful" (Elhissi 2017).

The mass median aerodynamic diameter (MMAD) is another measure used for determining aerosol particle size and is defined as the diameter where 50% of the aerosol mass lies above the specified size (Finlay and Darquenne 2020; Knap et al. 2023). A comprehensive evaluation of HDS, VMD, Dae, and MMAD provides a more robust understanding of particle/aerosol size dimensions, as each metric represents the particle/aerosol diameter from a unique perspective.

### 3.2.1.2 Particle Size Distribution
**Polydispersity Index (PDI)**

The PDI is a dimensionless measure of the uniformity of the particle size distribution in nanocarrier systems (Danaei et al. 2018). This value reflects the degree of nonuniformity (polydispersity) of the particles, with values below 0.05 indicating highly monodisperse samples and values above 0.7 suggesting broad size distributions unsuitable for DLS analysis (Danaei et al. 2018).

Another term used to measure the polydispersity of particles is the "span", which is used to indicate how particles are broadly distributed around the median size. Mathematically, span = [(90% undersize − 10% undersize)/VMD] (Elhissi et al. 2013).

**Geometric Standard Deviation (GSD)**

The GSD measures the spread of particle sizes around the median value, indicating the uniformity of the particle size distribution, with lower values reflecting greater uniformity (Finlay and Darquenne 2020; Knap et al. 2023). Particles having a GSD value of 1 are considered monodispersed, whereas those with a GSD greater than 1.2 are considered heterodispersed (Knap et al. 2023).

### 3.2.1.3 Particle Size/Mass Proportions
**Fine Particle Fraction (FPF) and Fine Particle Dose (FPD)**

The fine particle fraction (FPF) and fine particle dose (FPD) are both terms used to express how good aerosolized particles are at reaching the peripheral respiratory airways. FPF can be calculated as the percentage of particles with an aerodynamic diameter less than 5 μm (Knap et al. 2023). By contrast, the FPD represents the mass or weight of aerosolized particles with an aerodynamic diameter less than 5 μm and is derived from the total emitted dose of the formulation (Khan et al. 2023).

## 3.2.2    Zeta Potential

The zeta potential quantifies the electrostatic charge on the surface of liposomes and may also serve as an indicator of suspension stability under various environmental conditions (Németh et al. 2022). High zeta potential values ($> \pm 30$ mV) are generally associated with colloidal stability due to strong repulsion between the particles, which helps reduce the aggregation tendency (Németh et al. 2022; Öztürk et al. 2024; Pochapski et al. 2021). However, in biological contexts, particularly in drug delivery systems like liposomes, nanoparticles with high zeta potential values can lead to toxicity (Öztürk et al. 2024; Serrano-Lotina et al. 2023; Shao et al. 2015). For instance, cationic (positively charged) liposomes may interact strongly with negatively charged cell membranes, potentially causing membrane destabilization and increased permeability (Jin et al. 2021; Öztürk et al. 2024). Therefore, while a high absolute zeta potential enhances formulation stability, it is crucial to balance this with biocompatibility to minimize potential toxicological risks.

## 3.2.3    Surface Morphology

The physical appearance and structural integrity of proliposomes affect their subsequent drug release profiles and the overall stability of the resultant liposomes (Bibi et al. 2011). Microscopic techniques, such as scanning electron microscopy (SEM), are used for visualizing proliposome microparticles and can assist in the evaluation of the particle size and morphology of proliposome particles, including surface morphology of the particles, which can correlate with powder flowability (Bibi et al. 2011).

### 3.2.3.1    Proliposome Morphology Assessment

Some of the commonly employed methods for the morphology analysis of proliposomes include SEM, which is used to assess the external appearance of proliposome microparticles, including surface smoothness and flowability (Dallal Bashi et al. 2024; Khan et al. 2023; Patil-Gadhe and Pokharkar 2013). In one study, the particle morphology of proliposomes, which was examined via SEM, revealed distinct characteristics based on the formulation (Omer et al. 2018). The mannitol-based proliposome particles shown in Fig. 3.1 (coded F1–F5) presented morphological characteristics distinct from those of the lactose monohydrate (LMH)-based particles highlighted in Fig. 3.2 (coded F6–F10) (Omer et al. 2018). Notably, mannitol-based particles were consistently spherical regardless of the lipid-to-carrier ratio (Omer et al. 2018). By contrast, LMH-based proliposome particles exhibited an irregular shape, an apparently rough surface texture, and widely distributed particle sizes (Omer et al. 2018).

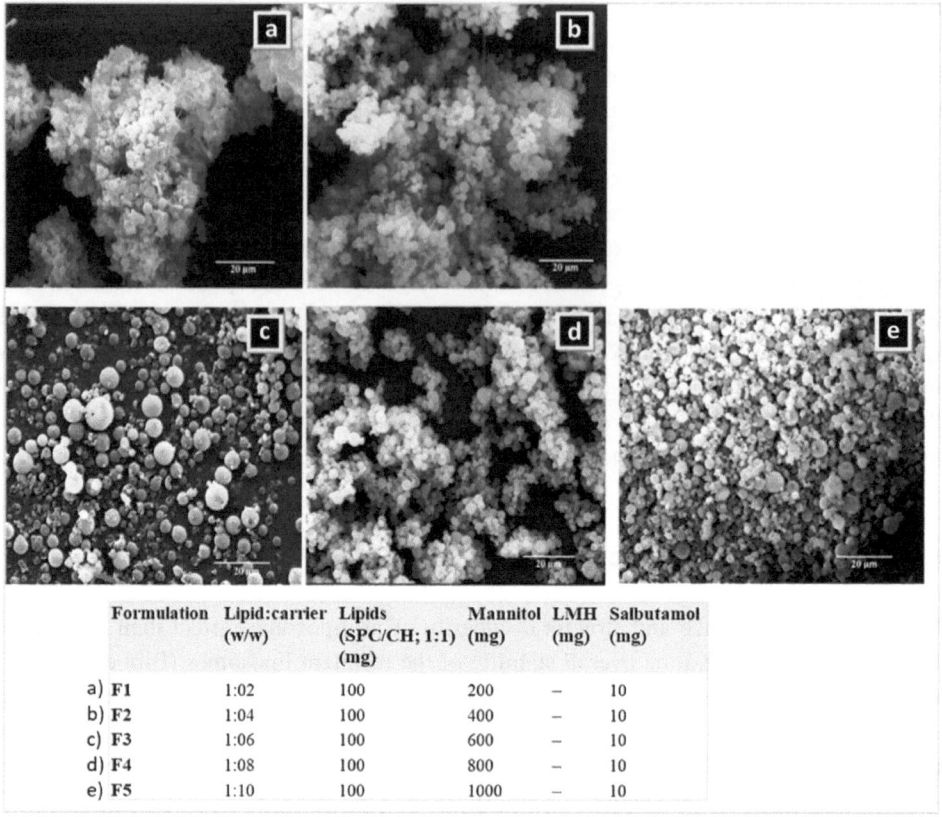

| Formulation | Lipid:carrier (w/w) | Lipids (SPC/CH; 1:1) (mg) | Mannitol (mg) | LMH (mg) | Salbutamol (mg) |
|---|---|---|---|---|---|
| a) F1 | 1:02 | 100 | 200 | – | 10 |
| b) F2 | 1:04 | 100 | 400 | – | 10 |
| c) F3 | 1:06 | 100 | 600 | – | 10 |
| d) F4 | 1:08 | 100 | 800 | – | 10 |
| e) F5 | 1:10 | 100 | 1000 | – | 10 |

**Fig. 3.1** SEM images of mannitol-based proliposomes (coded as F1–F5) showing the particle morphology. **a** F1, **b** F2, **c** F3, **d** F4, and **e** F5. Reproduced with permission from Omer et al. (2018), under the Creative Commons Attribution License (CC BY). © Springer Nature

### 3.2.3.2  Liposome Morphology Assessment

Additional morphological investigations of liposomes can be carried out after hydration of the proliposomes. A variety of microscopy techniques for visualizing liposome-based systems can be utilized, including light microscopy, fluorescence microscopy, confocal microscopy, and a range of electron microscopy methods, such as transmission electron microscopy (TEM), cryo-TEM, freeze-fracture electron microscopy, and environmental scanning electron microscopy (Bibi et al. 2011). In one study, Omer and co-workers revealed, through TEM, the formation of oligolamellar liposomes (OLVs) through manual hydration of mannitol-based proliposomes and the production of elongated worm-like bilayer liposomes along with liposome clusters from LMH-based proliposomes at a 1:6 w/w lipid-to-carrier ratio (Fig. 3.3). The study demonstrated that type of sugar carrier in

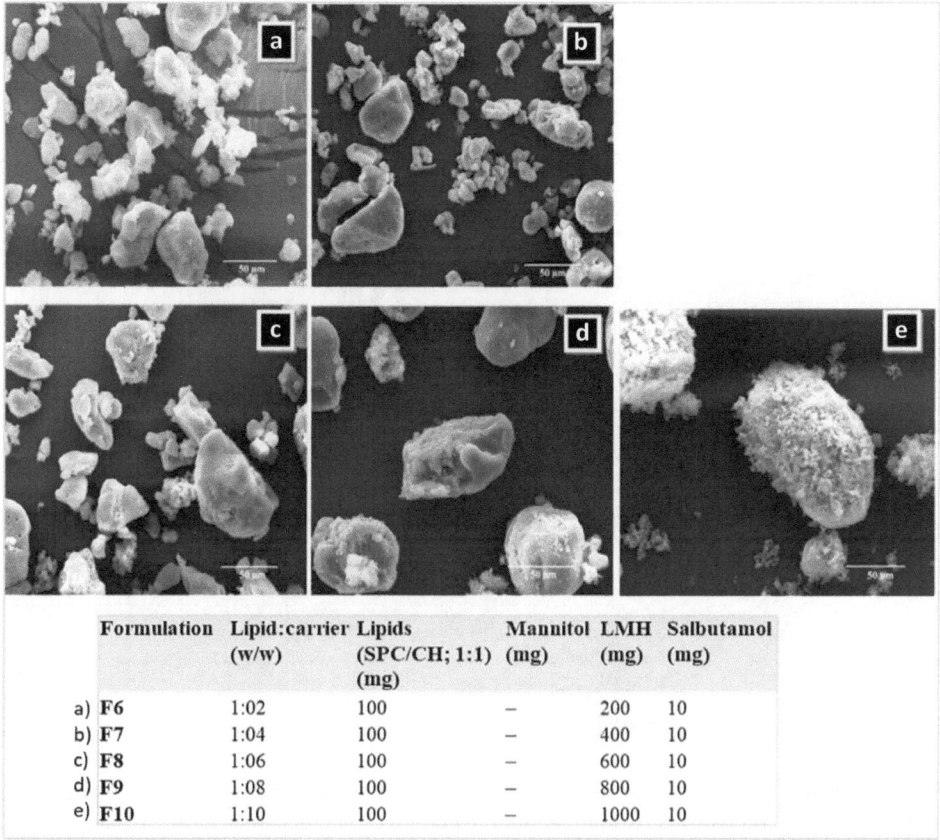

| Formulation | Lipid:carrier (w/w) | Lipids (SPC/CH; 1:1) (mg) | Mannitol (mg) | LMH (mg) | Salbutamol (mg) |
|---|---|---|---|---|---|
| a) F6 | 1:02 | 100 | – | 200 | 10 |
| b) F7 | 1:04 | 100 | – | 400 | 10 |
| c) F8 | 1:06 | 100 | – | 600 | 10 |
| d) F9 | 1:08 | 100 | – | 800 | 10 |
| e) F10 | 1:10 | 100 | – | 1000 | 10 |

**Fig. 3.2**  SEM images of LMH-based proliposomes (coded as F6–F10) showing particle morphology. **a** F6, **b** F7, **c** F8, **d** F9, and **e** F10. Reproduced with permission from Omer et al. (2018). © Springer Nature

proliposome formulations could influence the morphology of the resultant vesicles, which may affect the behavior of the vesicles in vivo.

## 3.2.4  Chemical Structure

Nuclear magnetic resonance (NMR), high-performance liquid chromatography-mass spectrometry (HPLC–MS), and Fourier transform infrared (FTIR) spectroscopy have been employed to study the chemical structure and molecular weights of drugs and excipients within proliposome-based formulations and identify their chemical bonds and interactions (Bonechi et al. 2021; Peleg-Shulman et al. 2001; Shi and Li 2023). Specifically, for proliposomes, these methods can help predict formulation stability and detect potential

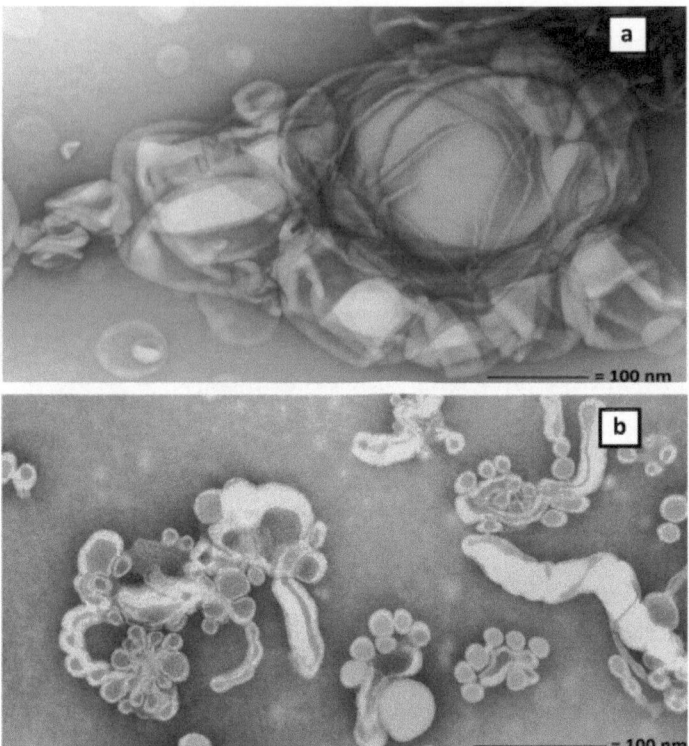

**Fig. 3.3** TEM images of **a** OLV liposomes generated upon manual hydration of mannitol-based proliposomes and **b** elongated worm-like bilayer liposomes and liposome clusters generated from LMH-based proliposomes using a 1:6 w/w lipid to carrier ratio. Reproduced with permission from Omer et al. (2018). © Springer Nature

chemical changes during proliposome preparation or storage, in order to ensure compatibility, efficacy, and stability. As highlighted in Fig. 3.4, FTIR studies were conducted to examine the interactions between the components of five proliposome formulations (Rojanarat et al. 2012b). The resulting spectra were analyzed to identify characteristic absorption bands, including plane vibrations, stretching vibrations, peaks, and alterations, which indicated chemical interactions between the excipients and the active pharmaceutical ingredient (API) (Rojanarat et al. 2012b).

## 3.2.5  Lipid Concentration Assay

The Stewart assay is a common analytical method used to determine the lipid concentration in proliposomal/liposomal formulation samples (Khan et al. 2021; Stewart 1980).

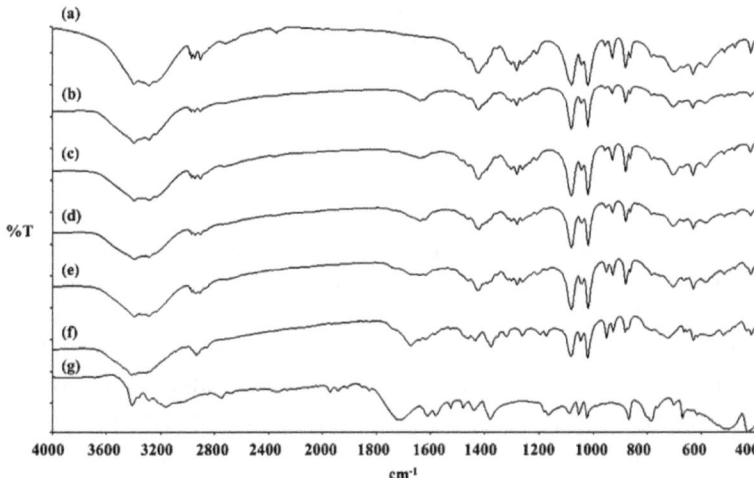

**Fig. 3.4** FTIR spectra of **a** porous mannitol, **b–f** pyrazinamide-proliposome formulations I–V (containing pyrazinamide 10, 20, 40, 60 and 80% w/w, respectively) and **g** pyrazinamide. Reproduced with permission from Rojanarat et al. (2012a). © Taylor & Francis

This assay is particularly useful for evaluating the efficiency of lipid coating by comparing the measured lipid content to the theoretical input (Khan et al. 2021; Stewart 1980). The Stewart assay has also been used to quantify the proportion of phospholipid deposited in FPF using inertial impaction (Elhissi and Taylor 2005). Generally, to analyze proliposomal preparations, a sample of a liposome suspension (hydrated from proliposomes) is dissolved in ethanol and incubated overnight at 90 °C. After adding chloroform and ammonium ferrothiocyanate solution, the mixture is vortexed followed by centrifugation in order to separate the layers (Khan et al. 2021). The lower chloroform layer, containing dissolved lipids, is aspirated and analyzed via a UV–Visible spectrophotometer at 488 nm to provide precise lipid quantification (Khan et al. 2021).

## 3.2.6  Diffraction Patterns

Powder X-ray diffraction (XRD) is used to analyze the diffraction patterns of proliposome powders, providing insights into the crystalline properties that can influence drug solubility and excipient affinity to the aqueous phase, and can be correlated with the subsequent release rates of the incorporated drug following hydration (Gomez et al. 2020; Rojanarat et al. 2011). For example, Omer et al. demonstrated the utility of XRD in distinguishing the crystalline properties of proliposome formulations influenced by different excipients

(Omer et al. 2018). This study explored how the choice of carriers, specifically mannitol and LMH, and the lipid-to-carrier ratio can affect the diffraction profiles and overall characteristics of proliposomes (Omer et al. 2018).

### 3.2.7 Thermal Analysis

Thermal analysis techniques are employed to study phase transitions and thermal stability, which are critical for storage and handling conditions. Differential scanning calorimetry (DSC) can assist in the assessment of thermal stability by evaluating enthalpy changes and drug entrapment in liposomes generated from proliposomes (Gomez et al. 2020; Rojanarat et al. 2011). In an investigation on the thermal behavior of traditional liposomes (made by thin-film hydration) versus liposomes generated from particulate-based or ethanol-based proliposomes, high-sensitivity DSC (HSDSC) was employed. Liposomes generated from ethanol-based proliposomes exhibited a main transition of higher enthalpy, a lower onset temperature, and very low incorporation of the antiasthma steroid beclometasone dipropionate (BDP), whereas thin-film liposomes and particulate-based proliposome methods generated liposomes, offering greater incorporation of BDP (1–2.5 mol%) but with differing thermal profiles qualitatively and quantitatively (Elhissi et al. 2006a). Thin-film liposomes showed phase separation at higher steroid concentrations (5 mol%), starting to form separate steroid domains within the liposome bilayers, which did not happen with vesicles generated from particulate-based proliposomes. This is likely to be attributed to the use of sucrose, which also reduced pretransition and main transition enthalpies (Elhissi et al. 2006a). In another study, DSC was used to determine whether BDP in the proliposome formulations was in a crystalline or amorphous state (Khan et al. 2015). The distinct endothermic peaks of individual proliposome components BDP, cholesterol, and sucrose are depicted in Fig. 3.5 (Khan et al. 2015). Combined analysis via light microscopy and DSC provided valuable insights into the behavior of BDP in both hydrated and dry proliposome formulations (Khan et al. 2015).

### 3.2.8 Flow Properties

Manufacture of free-flowing powder is essential for packaging, and dosing consistency, especially for solid dosage forms. Techniques such as bulk density, tapped density, and angle of repose (AOR) have widely been used to assess the cohesiveness and compressibility of proliposome powders (Khan et al. 2018, 2020; US Pharmacopeia 2015). Optimal flow properties prevent dosing errors and improve handling efficiency during production (Khan et al. 2018, 2020; US Pharmacopeia 2015). In the context of proliposomes, flowability affects the uniform coating of carrier particles and the subsequent release of the powder from the inhaler. A smooth, evenly coated proliposome carrier promotes efficient

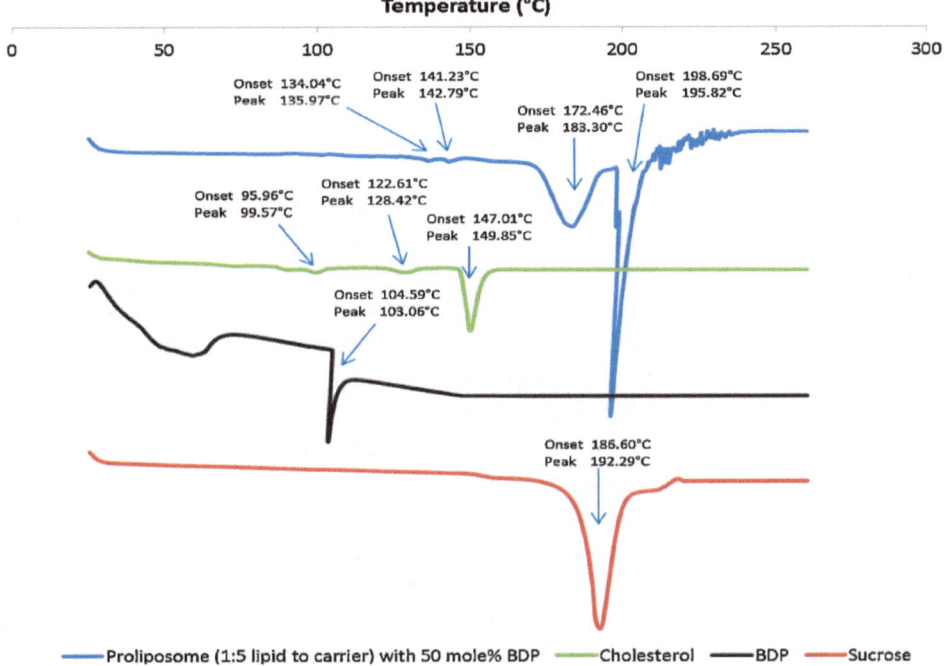

**Fig. 3.5** Superimposed DSC thermographs of BDP, cholesterol and sucrose compared with those of the proliposome with 50 mol% beclomethasone dipropionate, showing the absence of an endothermic peak of the drug in the proliposome formulation. Reproduced with permission from Khan et al. (2015). © Elsevier B.V.

hydration, generating liposomes of consistent size and enhancing drug delivery accuracy (Khan et al. 2018; US Pharmacopeia 2015).

In one of our studies, we have found that the emitted dose (ED) from a monodose powder inhaler loaded with spray-dried proliposomes, was consistently high across all prepared formulations, ranging from 77.46 to 94.59% (Omer et al. 2018). However, LMH-based proliposomes demonstrated greater deposition in the upper stage of a two-stage impinger, due to their larger particle sizes and irregular shapes, resulting in extremely poor FPF values (0–3.99%) and low powder flowability (Omer et al. 2018). Compared with LMH-based proliposome powders, mannitol-based microparticles, which are characterized by smaller, more spherical shapes, exhibited significantly better flowability and achieved a much higher FPF (2.79–52.14%) (Omer et al. 2018). In another study, flowability analysis aided in the selection of the optimal formulations for subsequent use in the delivery of a medication (Khan et al. 2020). Flowability studies identified three proliposomal preparations as the optimal formulations out of 27 tested compositions. These formulations demonstrated excellent AORs (e.g. 15.16° in one of the formulations). The high flowability correlated well with favorable compressibility indices (e.g. 14.56) (Khan et al. 2020).

### 3.2.9  Drug Encapsulation Efficiency (EE)

The drug encapsulation efficiency (EE) measures the effectiveness of drug incorporation into liposome structures, a crucial factor for optimizing drug delivery and providing sustained drug release (Knap et al. 2023). A high EE ensures that therapeutic goals are met with the maximum amount of drug involved in the provision of sustained release or targeted delivery (Knap et al. 2023). While high EE is desirable, it is important to avoid excessive lipid use because the entrapment efficiency depends on the lipid/aqueous phase ratio (Elhissi et al. 2006a). Therefore, optimizing carrier to lipid and drug to lipid ratios while maintaining high EE is essential (Elhissi et al. 2006a).

For antiasthma drugs such as BDP, significant differences in entrapment values were observed between proliposome-derived liposomes and traditional liposomes prepared via the thin-film method (Elhissi et al. 2006a). Liposomes generated from particulate-based proliposomes demonstrated optimal incorporation of 1–2.5 mol% BDP with apparently no formation of separate steroid domains or phase separation (Elhissi et al. 2006a). This stability is attributed to the inclusion of a sucrose carrier, which may prevent phase separation and maintain uniform mixing within the lipid bilayers (Elhissi et al. 2006a). By contrast, traditional thin-film-made liposomes exhibited similar optimal incorporation levels (1–2.5 mol%) for BDP but encountered phase separation at higher steroid concentrations, such as 5 mol% (Elhissi et al. 2006a). At these levels, distinct steroid domains formed within the bilayers, indicating less compatibility for higher drug doses than liposomes generated from particulate-based proliposomes (Elhissi et al. 2006a).

In another study, a novel "slurry method" was developed for the preparation of proliposome powders using soy phosphatidylcholine (SPC) and cholesterol (1:1), and incorporating BDP at 2 mol% of the total lipid phase (Khan et al. 2015). To confirm the feasibility of this new method, the vesicles produced were evaluated against those prepared via the conventional proliposome method and liposomes prepared by thin-film hydration, with a focus on vesicle size and drug entrapment efficiency (Khan et al. 2015). As illustrated in Fig. 3.6, the slurry method demonstrated superior BDP entrapment compared with both liposomes prepared using the traditional (feed-line) particulate-based proliposome approach and those produced through the thin-film hydration technique (Fig. 3.6) (Khan et al. 2015).

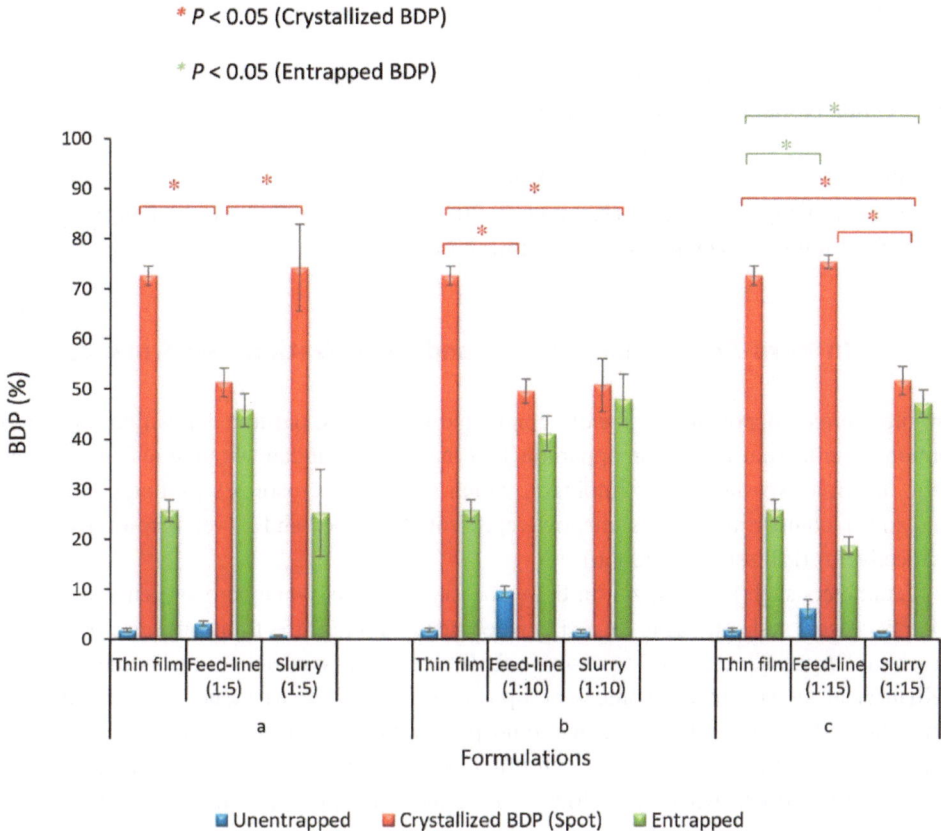

**Fig. 3.6** Entrapment efficiency of BDP in liposomes prepared the thin-film method and from proliposomes prepared via the feed-line and slurry methods. The formulations were prepared with the following lipid-to-carrier ratios: 1:5 w/w (**a**), 1:10 w/w (**b**) and 1:15 w/w (**c**) for the proliposome methods (feed-line and slurry-based). Reproduced with permission from Khan et al. (2015). © Elsevier B.V.

### 3.2.10 Loading Efficiency (LE)

Loading efficiency (LE) represents the proportion of the formulation's total mass that is composed of the drug, reflecting how much of the formulation is an active ingredient versus excipients (Knap et al. 2023). This is different from EE, which measures the proportion of the drug added during preparation that is successfully encapsulated (i.e., entrapped) within the delivery system (e.g., the liposomes) (Knap et al. 2023). While LE focuses on the formulation composition, EE assesses the effectiveness of the encapsulation process (Knap et al. 2023).

### 3.2.11 Emitted Dose (ED)

Emitted dose (%ED) represents the percentage of the drug dose fraction that is successfully emitted as aerosol/powder from the inhalation device out of the total amount initially loaded into the device (Knap et al. 2023). A higher ED indicates better device and formulation performance and minimal drug retention within the device, with a potentially maximized therapeutic benefit (Knap et al. 2023). ED is usually higher when the powder is made of uniform size and has good flowability.

### 3.2.12 In Vitro Aerosol Deposition and Nebulization Performance

In vitro aerosol deposition is usually conducted to evaluate particle deposition and distribution in an inertial impaction apparatus that may accommodate an aqueous environment (e.g., using a two-stage glass impinger, which separates aerosolized particles into upper and lower compartments to somehow represent the upper and lower respiratory tracts, respectively) (Elhissi et al. 2006b).

Parameters such as nebulization time (to "dryness", i.e. when aerosol generation completely ceases) and sputtering time (i.e. when nebulization starts to become erratic prior to reaching "dryness") are used to assess nebulizer efficiency in delivering the needed dose (Khan et al. 2021). In the context of liposome delivery, metrics such as aerosol mass/phospholipid output and aerosol mass/phospholipid output rates can also help in assessing nebulizer efficiency and how such efficiency is affected by formulation composition (Elhissi et al. 2006b; Khan et al. 2020). These measures provide scientific predictions for formulation eligibility to achieve optimal therapeutic delivery.

### 3.2.13 In Vitro and In Vivo Therapeutic Efficacy and Toxicity Studies

In vitro efficacy studies involve testing the proliposome formulation against target cells or microorganisms in a controlled laboratory setting. These studies help to assess the therapeutic potential of the formulation and identify the optimal dose for further investigations (Khan et al. 2020; Rojanarat et al. 2011). In vitro toxicity studies may involve assessing the effects of the formulation on cell viability, morphology, and function (Khan et al. 2020; Rojanarat et al. 2011).

In vivo therapeutic efficacy studies involve evaluating the proliposome formulations in animal models to assess the biological effects (Adel et al. 2021; Pokharkar et al. 2014). These studies provide insights into the pharmacokinetics, pharmacodynamics, and efficacy of the formulation in complex biological systems (Adel et al. 2021; Pokharkar et al. 2014). In vivo toxicity studies can be carried out concurrently to evaluate the adverse effects of the formulation in living organisms by examining various organs and tissues for possible signs of toxicity (Adel et al. 2021; Pokharkar et al. 2014).

## 3.3 Formulation Approaches Based on the Pulmonary Delivery System

Numerous proliposomal formulations for pulmonary delivery, which are primarily administered via diverse types and subtypes of inhalation devices, notably pMDIs, DPIs and medical nebulizers, have been explored in the literature. The formulations of proliposomes for delivery through these devices are discussed in details in the subsequent sections considering the research conducted by several research groups including our team.

### 3.3.1 Formulation of Proliposomes for pMDIs

#### 3.3.1.1 Formulation Characteristics of Proliposomes for pMDIs

Key research findings in proliposomes and liposome-based aerosol formulations are summarized in Table 3.2. The feasibility of formulating proliposomal preparations of chlorofluorocarbon (CFC) propellants for pulmonary delivery has been explored in a number of studies (Farr et al. 1987; Vyas et al. 2004, 2005; Vyas and Sakthivel 1994). For instance, Farr et al. demonstrated that solvent-based proliposomes comprising phospholipid dissolved in CFC blends could generate aerosol that is proposed to spontaneously form liposomes in situ within the respiratory tract upon contact with aqueous pulmonary fluids (Farr et al. 1987). Vyas and Sakthivel further advanced this concept by developing a similar CFC-based pressurized pack system for the delivery of the bronchodilator isoprenaline through proliposomes (Vyas and Sakthivel 1994). They investigated the spontaneous formation of liposomes within a single-stage glass impinger upon release of drug-phospholipid aerosols from a pressurized metered dose canister. Aliquots were taken for particle size analysis and drug entrapment efficiency studies (Vyas and Sakthivel 1994). They demonstrated that liposomes were formed in situ within the impinger and offered sustained drug release, with the lipid composition being the main factor to influence the drug release (Vyas and Sakthivel 1994). Figure 3.7 summarizes their formulation strategies and their corresponding outcomes (Vyas and Sakthivel 1994).

Notably, CFC propellants are excellent solvents for phospholipids, which allows the formulation of proliposome solutions within pMDI canisters (Farr et al. 1987; Vyas et al. 2004, 2005; Vyas and Sakthivel 1994). This process provided a promising means to deliver drugs in controlled-release formats, effectively delivering the drug in FPF while potentially minimizing systemic adverse effects. However, CFCs were found to have a detrimental effect on the ozone layer, leading to the Montreal Protocol, which phased out the use of CFCs in many countries (Elhissi 2017). This regulatory shift required

**Table 3.2** Summary of formulation details of the proliposome-based pMDIs explored in different studies

| Preparation method | API | Excipients | Propellant | pMDI device | Proposed indication | Author and year |
|---|---|---|---|---|---|---|
| Solution-phase pressurized packs with phospholipids in various CFC blends | None | Egg phosphatidylcholine | CFC | Pressurized pack aerosol device | Pulmonary drug delivery (general) | Farr et al. (1987) |
| Ether injection method and dissolution in CFC | Isoprenaline | Soybean lecithin, egg lecithin, dicetyl phosphate phosphatidyl ethanolamine, cholesterol | CFC | Pressurized pack aerosol device | Asthma management | Vyas and Sakthivel (1994) |
| Dissolution of microemulsions in dimethyl ether (DME) and propane | Lecithin | Lecithin, DME, propane | DME, propane | BK 632 actuator with a BK 357 valve | Pulmonary delivery of peptides | Sommerville et al. (2002) |
| Spray-drying with high-pressure homogenization | Budesonide | DSPC, calcium chloride dihydrate | HFA-134a | Aluminum canisters, Valois DF30/63 | Asthma, COPD | Tarara et al. (2004) |

(continued)

**Table 3.2** (continued)

| Preparation method | API | Excipients | Propellant | pMDI device | Proposed indication | Author and year |
|---|---|---|---|---|---|---|
| Phospholipids and cholesterol-based proliposomes modified by coating with alveolar macrophage specific ligands (MBSA and O-SAP) | Rifampicin | Egg PC, cholesterol, MBSA, O-SAP | CFC | Pressurized pack custom device | Tuberculosis | Vyas et al. (2004) |
| Phospholipids and cholesterol-based proliposome modified by coating with alveolar macrophage specific ligands (OPM and OPP) | Amphotericin B | Egg PC, cholesterol, OPM, OPP | CFC | Pressurized pack custom device | Pulmonary aspergillosis | Vyas et al. (2005) |

*API* active pharmaceutical ingredient, *CFC* chlorofluorocarbon, *COPD* chronic obstructive pulmonary disease, *DME* dimethyl ether, *DSPC* distearoylphosphatidylcholine, *HFA* hydrofluoroalkane, *MBSA* methylated bovine serum albumin, *MDI* metered-dose inhaler, *O-SAP* O-stearylamylopectin, *OPP* oleoyl palmitoyl phosphatidylethanolamine, *PC* phosphatidylcholine

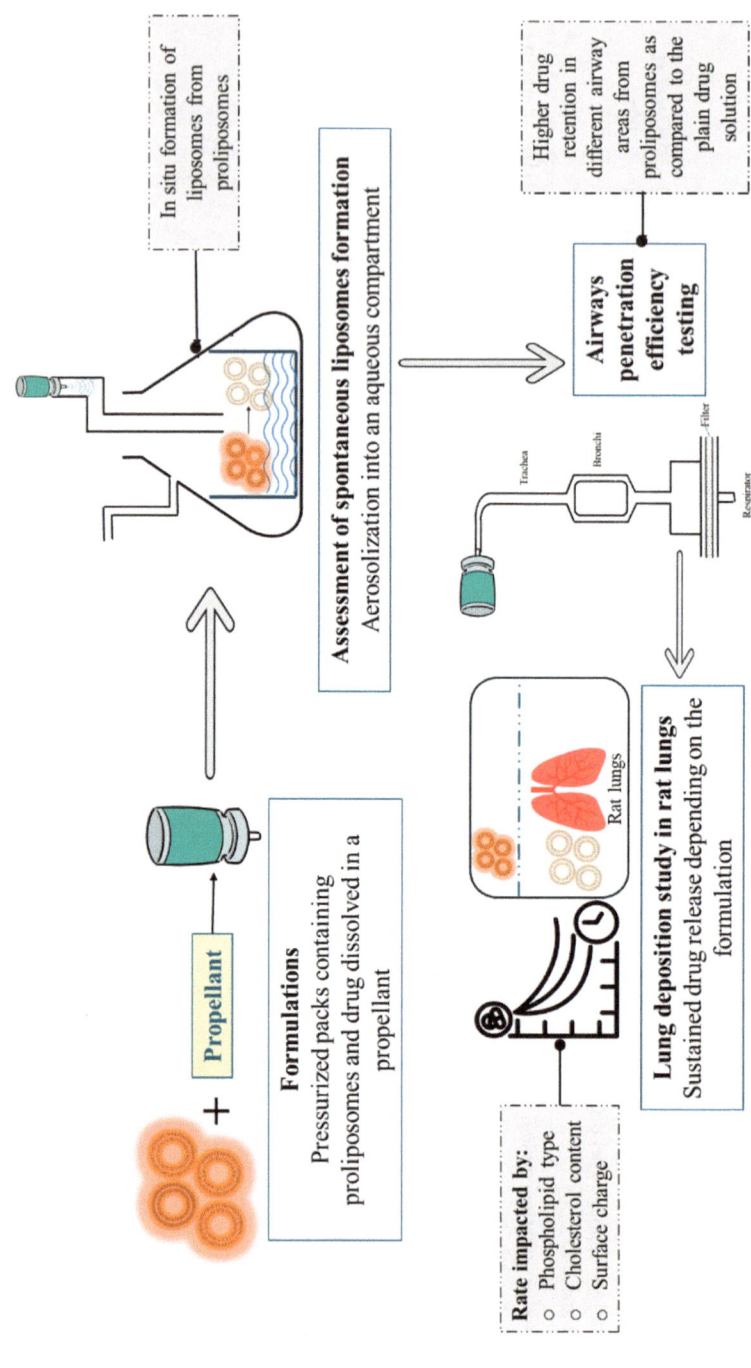

**Fig. 3.7** Schematic representation of pressurized pack formulations of proliposomes and their pulmonary delivery study outcomes

the replacement of CFCs with more environmentally friendly (i.e. non-ozone depleting) propellants, such as hydrofluoroalkane (HFA) propellants (Elhissi 2017).

Therefore, owing to the environmental concerns raised against CFCs, studies have begun to explore alternative pMDI formulations. For example, Sommerville et al. developed a lecithin-based microemulsion utilizing dimethyl ether (DME) and propane as propellants (Sommerville et al. 2002). These propellants play dual roles as solvents and dispersing agents, enabling the solubilization of polar compounds and peptides for pulmonary delivery (Sommerville et al. 2002). These formulations achieved FPFs of 50–70% and MMADs of 2.7–3.1 μm (Sommerville et al. 2002). Furthermore. Tarara et al. conducted a study in which the HFA propellant was used in lipid-coated proliposome formulations of budesonide (Tarara et al. 2004). The formulation utilized HFA-134a as the propellant and aluminum canisters equipped with Valois DF30/63 metering valves for delivery (Tarara et al. 2004). Despite HFA's limitations in solubilizing phospholipids, particles with MMADs of 3.2–3.4 μm were produced with stable suspensions over 6 months at 40 °C/75% relative humidity and 16 months at 25 °C/60% relative humidity (Tarara et al. 2004). This work highlighted how proliposome systems could somehow adapt to new regulatory and environmental constraints (Tarara et al. 2004).

While HFA propellants are ozone safe, they have different physicochemical properties than CFCs. The solubility of phospholipids in HFA propellants is poor; thus, research of proliposomes for delivery via pMDIs has significantly been diminished (Elhissi 2017). Thus, a greater focus towards alternative pulmonary drug delivery approaches involving proliposome formulations, such as those employing DPIs and medical nebulizers.

### 3.3.2 Formulation of Proliposome-Based DPIs

#### 3.3.2.1 Formulation Characteristics of DPI Proliposomes

The formulation of proliposome DPIs, in other words, particulate-based proliposomes for delivery in the form of aerosolized dry powders, involves careful selection of excipients and manufacturing and processing techniques to optimize drug delivery and stability.

As summarized in Table 3.3, multiple proliposome-based DPIs were investigated and found to incorporate a wide range of excipients. These excipients were selected to enhance formulation stability and aerosolization performance. Among the lipids used in the formulations are cholesterol, cholesterol sulfate, and stearylamine, which are integral for supporting the stability of liposomes, maximizing drug encapsulation and minimizing drug leakage from the vesicles following hydration of the proliposomes. Soya phosphatidylcholine (SPC), dipalmitoylphosphatidylcholine (DPPC), and hydrogenated soy phosphatidylcholine (HSPC) are among the phospholipids used in proliposome formulations, and are commonly mixed with an equimolar ratio of cholesterol (Table 3.3). Carriers such as lactose (monohydrates or microfines), mannitol, and sucrose, which are known to increase powder flow, dispersion, aerodynamic performance, and drug loading, are frequently used excipients in particulate-based proliposomes (Table 3.3). Surfactants, such as poloxamer 188 and TPGS, have also been utilized to reduce surface tension and improve particle dispersion during inhalation (Adel et al. 2021; Dallal Bashi et al. 2024). Trehalose is used as a stabilizer, preventing degradation during the spray drying process and storage (Aekwattanaphol et al. 2024; Dallal Bashi et al. 2024; Srichana et al. 2023; Wang et al. 2024b). Another key excipient used is L-leucine, which serves as a surface-active agent and is known for its role in reducing interparticle cohesiveness. This excipient may improve powder dispersibility and flowability, enhance aerodynamic properties, and ensure efficient lung deposition of DPI formulations (Adel et al. 2021; Aekwattanaphol et al. 2024; Dallal Bashi et al. 2024; Parhizkar et al. 2021; Patil-Gadhe and Pokharkar 2013; Srichana et al. 2023; Wang et al. 2024b).

Spray drying, which is known to facilitate the production of uniform, spherical particles with suitable aerodynamic properties for lung deposition, is the most prevalent method for preparing particulate-based DPI proliposomes (Table 3.3). For example, it allows the incorporation of functional excipients such as trehalose and L-leucine, which enhance particle stability and dispersion, as demonstrated in formulations of pretomanid for tuberculosis (Aekwattanaphol et al. 2024). Other methods used include ultrasonic spray freeze drying, the slurry method, jet milling and rotary evaporation (Parhizkar et al. 2021; Srichana et al. 2022; Wang et al. 2024b).

### 3.3.2.2 Proliposome Formulations for Different Indications Using Various DPIs

Various device types have been selected to deliver medications within proliposome-based DPI formulations based on their ability to optimize drug aerosolization, stability, and delivery efficiency. These include monodose capsule-based DPIs, specialized single-dose inhalers, multidose DPIs, and custom-made devices, which have been explored for diverse indications, including pulmonary tuberculosis, chronic respiratory infections, asthma, pulmonary arterial hypertension, lung cancer, pulmonary inflammation, pulmonary fungal infections, and *Pseudomonas aeruginosa* biofilms (Table 3.3).

Capsule-based DPIs, including the HandiHaler®, Aerolizer®, Neohaler®, Rotahaler®, Cyclohaler®, and RS01 low-resistance dry powder inhalers, were the main devices employed for the delivery of the proliposome-based DPI formulations (Table 3.3). These devices have shown consistent performance, delivering proliposomes with EE values often exceeding 50%, MMADs of 1.56–5.23 µm, and %ED frequently surpassing 87% for conditions such as tuberculosis, asthma, or chronic respiratory infections (Table 3.3). For example, Khan et al. developed spray-dried proliposome powder formulations of BDP, with formulations dispersed in water and ethanol (50:50% v/v) demonstrating significantly superior performance (Khan et al. 2023). These formulations achieved, on average, an FPF of 30.35%, FPD of 0.89 mg, and a respirable fraction of 85.90%, which were notably greater than those of the formulations dispersed in 100% ethanol (Khan et al. 2023). As demonstrated in Fig. 3.8, the study highlighted the critical role of dispersion

**Table 3.3** Summary of formulation details of the proliposome-based DPIs explored in different studies

| Method and number of formulations | API | Excipients | DPI device | Aerodynamic properties | Indication | Author and year |
|---|---|---|---|---|---|---|
| Co spray drying Number of formulations: 5 | Pretomanid | Trehalose, L-leucine, cholesterol and Phospholipon® | HandiHaler® | FPF: >50%, MMAD: <5 μm, HDS: 130–300 nm, %EE: 72.5%, %ED: >90%, GSD: 2.4–2.6 | Pulmonary tuberculosis | Aekwattanaphol et al. (2024) |
| Advanced co-spray drying in the closed mode Number of formulations: 3 | Fasudil monohydrochloride salt | DPPC, DPPG | Aerolizer®, Neohaler®, HandiHaler® | *Aerolizer®*: FPF: 58.3–60.3%, RF: 93.6–95.8%, ED: 94.8–97.2%, MMAD: 2.3–3.2 μm, GSD: 1.92–2.14 *Neohaler®*: FPF: 53.7–59.3%, RF: 85.8–93.8%, ED: 97.8–98.9%, MMAD: 2.8–3.0 μm, GSD: 1.87–2.40 *HandiHaler®*: FPF: 18.9–49.2%, RF: 33.4–77.9%, ED: 98.5–99.2%, MMAD: 3.3–11.6 μm, GSD: 2.30–3.42 | Pulmonary arterial hypertension | Ruiz et al. (2024) |

(continued)

**Table 3.3** (continued)

| Method and number of formulations | API | Excipients | DPI device | Aerodynamic properties | Indication | Author and year |
|---|---|---|---|---|---|---|
| Spray drying Number of formulations: 6 | Levofloxacin | Cholesterol, cholesterol sulfate, leucine, mannitol, trehalose, ammonium acetate | Cyclohaler® | FPF: 35–56%, MMAD: 2.02–2.37 μm, HDS: 500–1100 nm, %EE: 5.8–40.33%, %ED: 84–89% | Tuberculosis | Srichana et al. (2023) |
| Spray drying Number of formulations: 7 | Azithromycin | SPC, DDAB, TPGS, trehalose, L-leucine | Aerolizer® | *Formulations containing azithromycin: FPF: 19.8%, MMAD: 5.23 μm, HDS: 104.8–105 nm (after reconstitution), %EE: 74.9–87.3%, %ED: 87.3%, GSD: 2.17* | Chronic respiratory infections | Dallal Bashi et al. (2024) |
| Ultrasonic spray freeze drying (USFD) Number of formulations: 6 | Ciprofloxacin, colistin | HSPC, EPC, cholesterol, DSPE-PEG2000, sucrose, mannitol, trehalose, L-leucine, DTPA | Free-breath® single-dose capsule inhaler | *For the optimal formulation (F6): FPF: 44.44%, MMAD: 4.27 μm, %EE: 84.21, %ED: 99.3* | *P. aeruginosa* biofilm | Wang et al. (2024b) |

(continued)

**Table 3.3** (continued)

| Method and number of formulations | API | Excipients | DPI device | Aerodynamic properties | Indication | Author and year |
|---|---|---|---|---|---|---|
| Spray drying Number of formulations: 10 | Beclomethasone dipropionate | SPC, LMH, LMF, lactose 003, lactose 220, lactose 300 | RS01 low-resistance dry powder inhaler | FPF: 2.68–33.42%, MMAD: 2.98–14.05 μm, %EE: 78.16–96.89%, %ED: 78.55–93.64%, %RF: 19.12–89.47% | Asthma | Khan et al. (2023) |
| Slurry method and spray drying Number of formulations: 40 (20 with porous carriers and 20 with nonporous carriers) | Rifampin | Lactose, mannitol, sucrose, citric acid, cholesterol, L-α-lecithin | Aerolizer® | *For the optimized formulations (P11 and P18):* FPF: 0.71–9.17%, MMAD: 5.25–6.21 μm, %EE: 47.6–60.3%, ED: 0.95–1.63 mg, GSD: 2.36–5.42 | Pulmonary tuberculosis | Parhizkar et al. (2021) |
| Jet milling/ rotary evaporation Number of formulations: 8 | Isoniazid, rifampin | Cholesterol, cholesterol sulfate | HandiHaler® | *For isoniazid formulations:* FPF: 0.69–57.8%; MMAD: 2.06–2.61 μm; %ED: 70.1–99.4%; GSD: 1.05–2.89 *For rifampin formulations:* FPF: 2.6–57.4%; MMAD: 1.87–2.38 μm; %ED: 69.0–92.0%; GSD: 1.14–3.08 | Pulmonary tuberculosis | Srichana et al. (2022) |

(continued)

**Table 3.3** (continued)

| Method and number of formulations | API | Excipients | DPI device | Aerodynamic properties | Indication | Author and year |
|---|---|---|---|---|---|---|
| Nanospray drying Number of formulations: 26 | Curcumin | Lecithin, cholesterol, stearylamine, poloxamer 188, hydroxypropyl-beta-cyclodextrin (HPβCD) | Aerolizer® | *For improved SD/P4 formulation:* FPF: 54.35%; MMAD: 2.10 μm; %EE: 93.40%; %ED: 98.40% *For improved SD/P6 formulation:* FPF: 29.33%; MMAD: 3.18 μm; %EE: 93.40%; %ED: 79.96% | Lung cancer and inflammation | Adel et al. (2021) |
| Spray drying/ co spray drying Number of formulations: 6 | Amphotericin B | DPPC, DPPG | HandiHaler® | FPF: 13.0–49.9%, MMAD: 2.0–12.1 μm, HDS: not explicitly provided, %EE: 1.20–1.67%, %ED: 66.6–89.7%, %RF: 32.4–93.6%, GSD: 1.8–3.5 | Pulmonary fungal infections | Gomez et al. (2020) |
| Spray drying Number of formulations: 10 | Salbutamol sulphate | SPC, cholesterol, LMH, mannitol | MIAT monodose powder inhaler | *Mannitol-based formulations:* FPF: 2.79–52.14%; VMD: 3.38–6.01 μm; %EE: 16.70–22.57% *LMH-based formulations:* FPF: 0–3.99%; VMD: 3.23–5.96 μm; %EE: 15.46–37.76% | Asthma | Omer et al. (2018) |

(continued)

**Table 3.3**  (continued)

| Method and number of formulations | API | Excipients | DPI device | Aerodynamic properties | Indication | Author and year |
|---|---|---|---|---|---|---|
| Spray drying Number of formulations: 9 | Rifapentine | HSPC, cholesterol, stearyl amine, stearic acid, L-leucine, lactose (respitose SV010) | Rotahaler® | FPF: 60–92%, MMAD: 1.56–5.26 µm, %EE: 22.71–72.08%, %ED: >90% | Pulmonary tuberculosis | Patil-Gadhe and Pokharkar (2013) |
| Spray drying Number of formulations: 5 | Pyrazinamide | SPC, cholesterol, porous mannitol, ammonium carbonate | A custom-made delivery device | FPF: 19.4–29.0%, MMAD: 4.26–4.39 µm, HDS: 216–516 nm, %EE: 26.7–44.6%, %ED: 80.2–97.8% | Pulmonary tuberculosis | Rojanarat et al. (2012a) |
| Spray drying Number of formulations: 5 | Levofloxacin | SPC, cholesterol, porous mannitol, ammonium carbonate | A custom-made delivery device | FPF: 13.5–38.1%, MMAD: 4.15–4.44 µm, HDS: 466.8–1005.3 nm, %EE: 14.9–23.9%, %ED: 76.3–91.3% | Pulmonary tuberculosis | Rojanarat et al. (2012b) |
| Spray drying Number of formulations: 5 | Isoniazid | SPC, cholesterol, mannitol | Diskhaler® | FPF: 15–35%, MMAD: 2.99–4.92 µm, %EE: 17–30%, %ED: 91–95% | Pulmonary tuberculosis | Rojanarat et al. (2011) |

*API* active pharmaceutical ingredient, *DDAB* dimethyl dioctadecyl ammonium bromide, *DPI* dry powder inhaler, *DPPC* dipalmitoylphosphatidylcholine, *DPPG* dipalmitoylphosphatidylglycerol, *DSPE-PEG2000* distearoylphosphatidylethanolamine-polyethylene glycol 2000, *DTPA* diethylenetriaminepentaacetic acid, *EPC* egg phosphatidylcholine, *FPF* fine particle fraction, *GSD* geometric standard deviation, *HDS* hydrodynamic size, *HSPC* hydrogenated soy phosphatidylcholine, *LMH* lactose monohydrate, *LMF* lactose microfine, *MMAD* mass median aerodynamic diameter, *SPC* soybean phosphatidylcholine, *TPGS* D-alpha-tocopheryl polyethylene glycol succinate, *%EE* percent entrapment efficiency, *%ED* percent emitted dose, *VMD* volume median diameter media, irrespective of the carrier type employed, in impacting formulation properties for pulmonary drug delivery (Khan et al. 2023).

Specialized single-dose inhalers, including the Free-breath® single-dose capsule inhaler and the MIAT Monodose powder inhaler, are designed for high precision in single-dose administration (Omer et al. 2018; Wang et al. 2024b). These devices have delivered proliposomes with FPF values up to 52.14%, MMADs in the range of 3.38–6.01 µm, and %EDs reaching 99.3%, which can potentially ensure efficient pulmonary delivery for indications such as asthma (Omer et al. 2018) and infectious diseases (against *Pseudomonas aeruginosa* biofilms) (Wang et al. 2024a, b). For example, spray-dried proliposome formulations of salbutamol sulfate were developed using LMH or mannitol as carriers and various lipid-to-carrier ratios (1:2–1:10) (Omer et al. 2018). Mannitol-based proliposomes were spherical and had higher fine particle fraction (FPF of approx. 52.6%), whereas LMH-based proliposomes were irregular but showed higher drug entrapment efficiency (EE, 37.76% at a 1:2 ratio) (Omer et al. 2018). Upon hydration, mannitol formulations formed oligolamellar vesicles, whereas LMH generated 'worm-like' structures (Fig. 3.3) (Omer et al. 2018). The ED was relatively high for all formulations (77.46–94.59%) (Omer et al. 2018). The study demonstrated the potential of spray-dried proliposomes for efficient pulmonary drug delivery, with performance dependent on formulation (Omer et al. 2018). Figure 3.9 provides a representative summary of the key findings reported in the study conducted by Omer et al. (2018).

Diskhaler®, a multidose DPI device, has also been used for delivering powdered proliposome formulations encapsulating isoniazid (Rojanarat et al. 2011). FPF values have

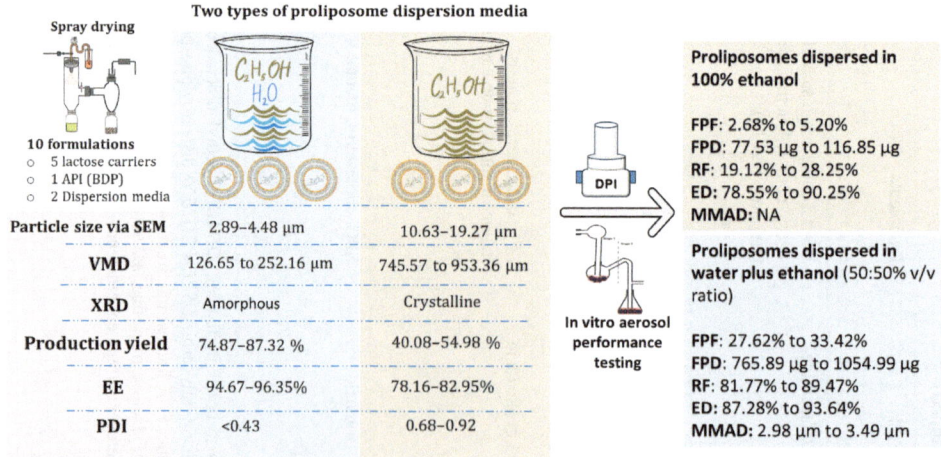

**Fig. 3.8** The impact of the dispersion medium on the proliposome formulation outcomes demonstrated the superiority of the combination of water and ethanol over ethanol alone, regardless of the carrier type, as reported by Khan et al. (2023)

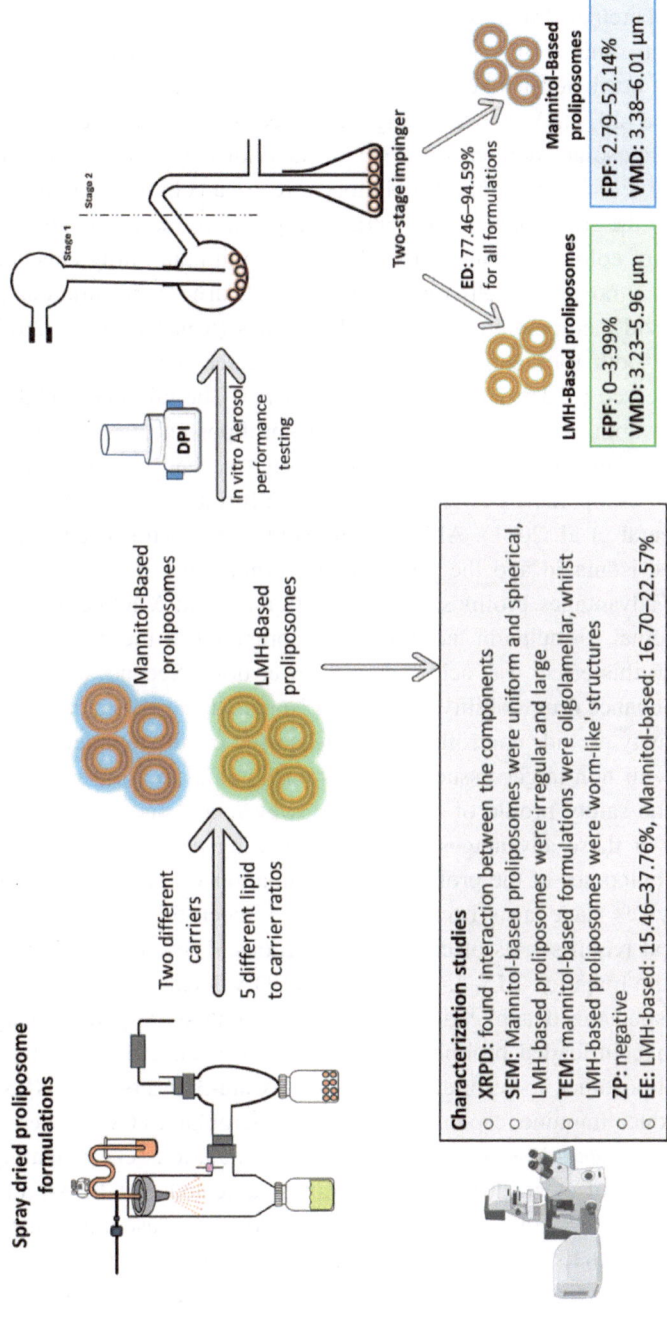

**Fig. 3.9** Representation of proliposome DPI formulations and their pulmonary delivery study outcomes, as reported by Omer et al. (2018)

been reported to be in the range of 15–35%, MMAD of 2.99–4.92 µm, and %ED of 91–95% (Rojanarat et al. 2011). These parameters are critical for effective pulmonary deposition of antitubercular drugs. Remarkably, these proliposome formulations demonstrated greater safety and antimycobacterial activity than free isoniazid when tested in vitro, which was attributed to their ability to increase intracellular drug uptake ($p < 0.05$) (Rojanarat et al. 2011). As shown in Fig. 3.10, free isoniazid and selected mannitol-based isoniazid-proliposomes demonstrated similar activities against M. bovis (Rojanarat et al. 2011). On day 1 of incubation, no inhibitory effect was observed at any concentration. By day 3, both the free isoniazid and optimized proliposome formulations achieved complete clearance of colony-forming units (CFUs) at a minimum drug concentration of 0.4 µg/mL. By day 7, no viable cells remained across all drug concentrations for either free drugs or isoniazid incorporated into the proliposomes (Rojanarat et al. 2011). These results indicate that both formulations effectively eradicated the bacilli by day 3, with greater efficacy observed by day 7 (Rojanarat et al. 2011). The minimum inhibitory concentration (MIC) for both free isoniazid and isoniazid-proliposome was 0.4 µg/mL on day 3 and less than 0.2 µg/mL on day 7. These findings suggest that the activity of isoniazid against M. bovis was comparable between the pure drug formulation and the proliposome formulation (Rojanarat et al. 2011). Although the study demonstrated comparable MIC values for both free isoniazid and the proliposome formulation, it is important to consider the potential advantages proliposomes may offer beyond MIC equivalence. First, the tested proliposomal formulation had a low EE compared to the other formulations, which indicates that this effect was achieved at a lower dose. Second, drug delivery via proliposomes can enhance drug stability and protect isoniazid from degradation. Additionally, proliposomes may provide controlled drug release and dosing regimens. By reducing direct interaction with non-target tissues, proliposomes could also lower systemic toxicity and improve the safety profile of isoniazid. However, the study could benefit from further exploration of these advantages, such as comparing pharmacokinetics and other pharmacodynamic outcomes of the proliposome formulation versus free isoniazid.

Customized devices have also been developed to meet the specific aerodynamic and physicochemical requirements of certain formulations. These devices have achieved MMAD values in the range of 4.15–4.44 µm, and %ED values reaching 97.8%, making them suitable for specialized applications such as tuberculosis targeting using pyrazinamide and levofloxacin-loaded proliposomes (Rojanarat et al. 2012a, b). Remarkably, the toxicity of proliposomes to respiratory cell lines (Calu-3, A549, and NR8383) and their potential to induce immune responses in alveolar macrophages (AMs) were assessed, along with in vivo repeated-dose toxicity in rats for formulations encapsulating pyrazinamide (PZA) (Rojanarat et al. 2012a). The findings indicated that PZA-proliposomes were less toxic to respiratory cells and did not trigger AMs to release inflammatory mediators such as interleukin-1β, tumor necrosis factor-α, and nitric oxide at toxic levels. Additionally, no renal or liver toxicity was observed in rats. Phagocytosis of dry PZA

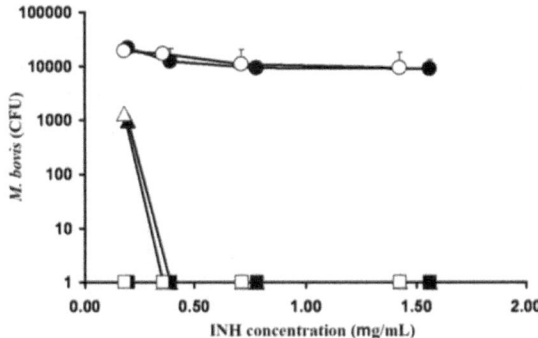

**Fig. 3.10** The reduction of *M. bovis* (CFU) after incubation with free isoniazid (INH) (dark symbol) and a selected isoniazid-proliposome formulation (blank symbol) on day 1 (●, ○), day 3 (▲, △) and day 7 (■, □) (mean ± SD, n ≥ 3). Figure reused from Rojanarat et al. (2011), under the Creative Commons Attribution License (CC BY 3.0)

powder particles by macrophages was evaluated, and the results revealed that phagocytosis occurred within 2 min, as there was no significant difference between the findings at 2 and 30 min (Fig. 3.11) (Rojanarat et al. 2012a). This process was evidenced not only by the increased intensity of the red fluorescence signal from the Lumidot® 640 stain, but also by the enlargement and shape changes of the NR8383 cells (Rojanarat et al. 2012a). These findings confirmed that PZA-liposomes derived from PZA-proliposomes can be phagocytosed by NR8383 cells (Rojanarat et al. 2012a). When the same investigations were replicated on proliposomal formulations incorporating levofloxacin, comparable findings were concluded (Rojanarat et al. 2012b).

In summary, proliposome-based DPIs offer a useful approach for treating a wide range of respiratory conditions. The choice of method, carriers, and excipients plays a pivotal role in optimizing aerodynamic properties and therapeutic outcomes. The successful application of these DPIs across tuberculosis, asthma, lung infections, and other pulmonary diseases highlights the potential of proliposome DPIs as noninvasive, effective drug delivery systems. Further advancements in formulation techniques and tailored excipient combinations can broaden their applicability and clinical potentials. Proliposome DPIs must also retain stability under various storage conditions, which is a matter that should be further investigated in the future.

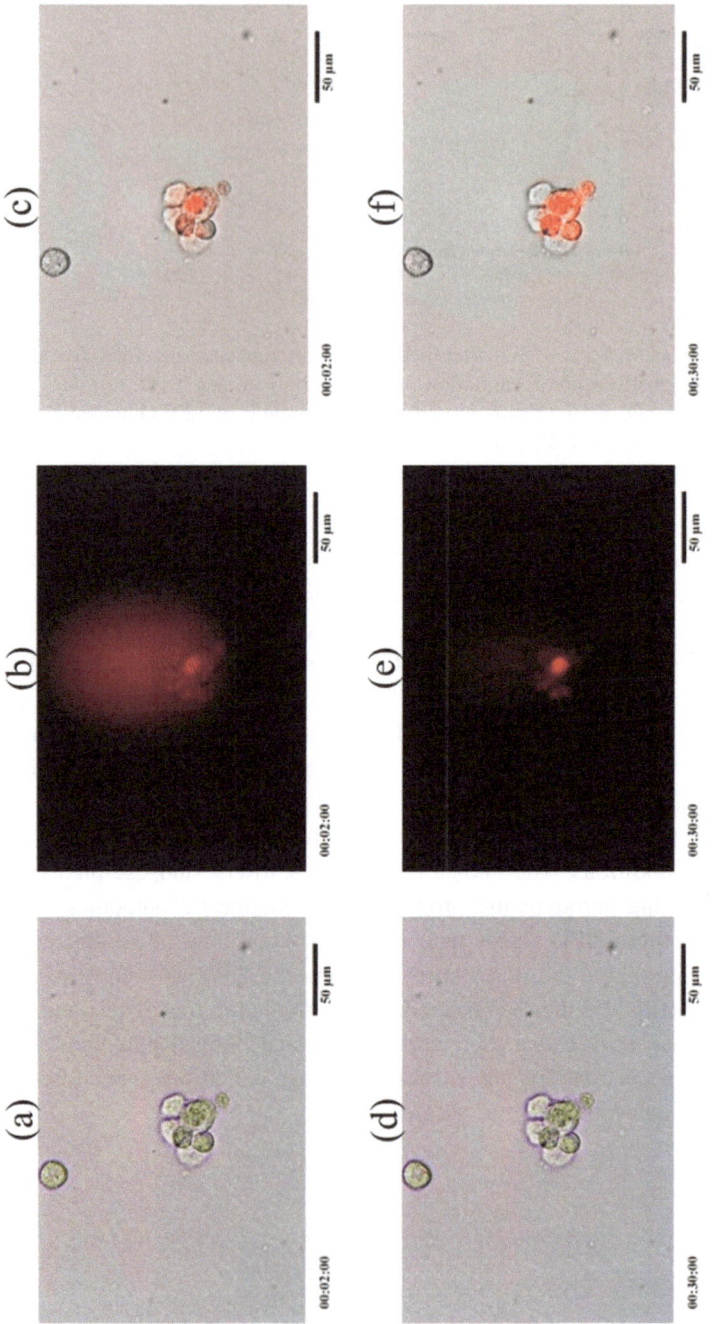

**Fig. 3.11** Phagocytosis of NR8383 cells incubated with reconstituted PZA-proliposome formulation I stained with Lumidot® 640 at **a–c** 2 and **d–f** 30 min in different modes; **a, d** brightfield image, **b, e** fluorescence image and **c, f** overlay image (bar = 50 μm). Reproduced with permission from Rojanarat et al. (2012a). © Taylor & Francis

### 3.3.3    Formulation of Proliposomes or Liposomes Generated from Proliposomes for Nebulization

#### 3.3.3.1    Formulation Characteristics of Proliposomes for Nebulization

Like other pulmonary delivery devices, the formulation of proliposomes for nebulization involves selection of lipids, carriers, and excipients in optimal proportions to ensure physical stability of liposomes, maximize drug retention within the vesicles, and achieving effective nebulization performance, including the achievement of high FPF values.

Lipid composition of proliposomes plays a crucial role in ensuring the stability and efficient retention of drug encapsulation. As outlined in Table 3.4, common lipids utilized in proliposome formulations include SPC, DPPC, and cholesterol, which help form physically stable liposomes during nebulization. For example, SPC was used in the preparation of particulate-based proliposomes to generate nebulizable liposomes, where FPF of aerosol droplets was 79.85–84.76%, and VMD was 2.45–2.50 µm (Elhissi et al. 2012).

Excipients such as mannitol, sucrose, and sorbitol are frequently included to serve as lipid carriers and facilitate formulation hydration into liposomes and aerosolization from nebulizers (Table 3.4). For example, mannitol was used in formulations of BDP-loaded proliposomes and was nebulized via a jet-nebulizer (Khan et al. 2021). Chitosan chloride has also been used in some formulations to enhance biocompatibility, biodegradability, and mucoadhesiveness (Albasarah et al. 2010).

A range of explored proliposome formulations have been prepared via methods such as film deposition via the carrier technique, slurry method, spray drying, and rotary evaporation to produce particulate-based proliposomes, each of which produced formulations with distinct properties (Table 3.4). For example, the film deposition method has been used to produce proliposomes with high EEs, such as favipiravir-loaded formulations, which achieved EE values ranging from 86.25 to 92.36% (Shaik et al. 2022).

#### 3.3.3.2    Proliposome Formulations for Nebulization

To explore the use of proliposomal formulations within medical nebulizers, to the best of our knowledge, Elhissi and Taylor were the first who investigated particulate-based proliposomes as a formulation capable of forming isotonic liposome dispersions in situ within air-jet (Pari LC Plus), ultrasonic (Liberty), and vibrating-mesh (Omron NE-U22) nebulizers (Elhissi and Taylor 2005). This was conducted by utilizing the energy/mixing provided during nebulization to convert the proliposomes within the nebulizer reservoirs into liposomes for delivery to a two-stage impinger (TSI) (Elhissi and Taylor 2005). Both the air-jet and vibrating-mesh nebulizers efficiently delivered liposomes to the second stage of the TSI, indicating their potential for effective delivery to the peripheral airways (Elhissi and Taylor 2005). However, as shown in Fig. 3.12, the phospholipid output from these nebulizers was lower than the total mass output, suggesting some accumulation of liposomes within the nebulizer reservoir during nebulization. By contrast, the ultrasonic nebulizer delivered less than 6% of the phospholipids to the impinger, indicating

**Table 3.4** Summary of formulation details of proliposomal preparations delivered via medical nebulizers

| Preparation method | API | Excipients | Nebulizer name | Formulation properties | Nebulization performance | Indication | Author and year |
|---|---|---|---|---|---|---|---|
| Film deposition on carrier method Number of formulations: 6 | Favipiravir | DPPC, SPC, cholesterol, mannitol | Optimo-NEB-102 model, compressor nebulizer | Mean particle size: 373.2 nm, %RF: 8.14–12.31%, %EE: 86.25–92.36% | Nebulization time: 12.95–14.3 min, aerosol output rate: 0.17–0.22 mg/min, aerosol mass output: 44.31–54.28%, respirable fraction: 8.14–12.31% | Viral infection | Shaik et al. (2022) |
| Slurry method Number of formulations: 8 (4 powders, 4 tablets) | Beclometasone dipropionate | SPC, cholesterol, sorbitol, mannitol | Jet nebulizer (Pari-LC Sprint) | Powder: VMD: 5.49–5.68 µm, %EE: 37.30–52.78% Tablet: VMD: 5.39–5.85 µm, %EE: 40.08–54.22% | Powder: nebulization time: 24–28 min, aerosol output rate: 160–170 mg/min, aerosol mass output: >80% Tablet: nebulization time: 23–31 min, aerosol output rate: 140.92–180.20 mg/min, aerosol mass output: >80% | Asthma | Khan et al. (2021) |

(continued)

**Table 3.4** (continued)

| Preparation method | API | Excipients | Nebulizer name | Formulation properties | Nebulization performance | Indication | Author and year |
|---|---|---|---|---|---|---|---|
| Slurry method and spray drying Number of formulations: 27 powders and 27 tablets | Paclitaxel | SPC, HSPC, DMPC, cholesterol, LMH, MCC, starch | Ultrasonic nebulizer (Uniclife Healthcare), vibrating mesh nebulizer (Omron NE-U22) | *For the optimal formulation (F3 powder):* particle size: 5.35 ± 0.76 µm, %EE: 95.45 ± 2.78% | *Vibrating mesh nebulizer (F3 tablet):* nebulization time: 15.73 ± 1.23 min, aerosol output rate: 306.72 mg/min, aerosol mass output: 98.56% *Ultrasonic nebulizer (F3 tablet):* nebulization time: 8.75 ± 0.86 min, aerosol output rate: 421.06 mg/min, aerosol mass output: 84.03% | Cancer | Khan et al. (2020) |
| Ethanol-based proliposome method Number of formulations: 2 | Salbutamol sulphate | SPC, cholesterol | Aeroneb Pro vibrating-mesh nebulizer | FPF (of aerosol droplets): 57.85%, VMD (of aerosol droplets): 3.44 ± 0.05 µm | Aerosol mass output: 90% | Asthma | Elhissi et al. (2013) |
| Particulate-based proliposome method (rotary evaporation) Number of formulations: 3 | None | SPC, EPC, cholestrol, sucrose | Pari LC Plus jet nebulizer | FPF (of aerosol droplets after 10 min): 79.85–84.76%, VMD (of aerosol droplets after 10 min): 2.45–2.50 µm | Aerosol mass output: >80% | Pulmonary delivery (general) | Elhissi et al. (2012) |

(continued)

**Table 3.4** (continued)

| Preparation method | API | Excipients | Nebulizer name | Formulation properties | Nebulization performance | Indication | Author and year |
|---|---|---|---|---|---|---|---|
| Ethanol-based proliposome method Number of formulations: 2 | Salbutamol sulfate | SPC, cholesterol, NaCl, sucrose | Aeroneb Go vibrating-mesh | FPF: 57.44%, VMD: 19.38 μm (main reservoir), 3.53 μm (lower reservoir), 4.27 μm (upper stage of the impinger), 6.72 μm (lower stage of the impinger), %RF: (main reservoir), 78% (lower reservoir), 22% (upper stage of the impinger), 18%, (lower stage of the impinger): 12% | Aerosol mass output: 81.92% in NaCl and 32.47% in sucrose | Asthma | Elhissi et al. (2011a) |
| Particulate-based proliposome method (rotary evaporation) Number of formulations: 2 | Beclometasone dipropionate | DMPC, sucrose | Pari LC Plus jet nebulizer | VMD: 6.95 μm (nebulizer)/ 4.30 μm (impinger) | Aerosol mass output: ~80% | Asthma | Elhissi et al. (2011b) |

(continued)

**Table 3.4** (continued)

| Preparation method | API | Excipients | Nebulizer name | Formulation properties | Nebulization performance | Indication | Author and year |
|---|---|---|---|---|---|---|---|
| Ethanol-based proliposome method. Number of formulations: 4 | Amphotericin B | SPC, chitosan chloride | Pari LC Sprint air-jet nebulizer | VMD: 2.98–5.08 µm, hydrodynamic diameter: 172.2–211.4 nm, %EE: 66.7–79.5% | Aerosol droplet size: 2.5–3.3 mm, total aerosol mass output: 93.1–96.1%, AmB deposition in lower stage: 58.4–61.3% | Pulmonary fungal infections | Albasarah et al. (2010) |
| Ethanol-based proliposome method. Number of formulations: 3 | Salbutamol sulfate | SPC, cholesterol, NaCl, sucrose | Pari LC Plus jet nebulizer, Aeroneb Pro (4 and 8 µm mesh), Omron NE U22 vibrating-mesh nebulizer | VMD *Pari*: 5.20 µm (reservoir), 3.60 µm (upper stage), 2.83 µm (lower stage) *Aeroneb (4 µm mesh)*: 6.42 µm (reservoir), 3.19 µm (upper stage), 2.50 µm (lower stage) *Omron*: 4.75 µm (reservoir), 3.19 µm (upper stage), 2.84 µm (lower stage) | Mass output (%): Pari: 85.5–93.18%; Aeroneb (4 µm): 60.13–89.94%; Aeroneb (8 µm): 64.29–82.95%; Omron: 97.81–99.93% Mass output rate (mg/min): *Pari*: 191.02–177.06; *Aeroneb (4 µm)*: 221.95–318.54; *Aeroneb Pro (8 µm)*: 251.20–478.68; *Omron*: 88.12–174.48 | Pulmonary delivery | Elhissi et al. (2006a, b) |

(continued)

**Table 3.4** (continued)

| Preparation method | API | Excipients | Nebulizer name | Formulation properties | Nebulization performance | Indication | Author and year |
|---|---|---|---|---|---|---|---|
| Particulate-based proliposome method (rotary evaporation) Number of formulation: 1 | None | SPC, cholesterol, sucrose | Pari LC Plus (jet), Liberty (ultrasonic), Omron NE U22 (vibrating-mesh) | VMD (µm) after nebulization: *Pari*: 9.04 (1 min), 5.42 (5 min); *Liberty*: 64.29 (1 min), 8.75 (5 min); *Omron*: 49.93 (1 min), 12.63 (5 min) | Mass output (total, %): Pari: 87, Liberty: 55, Omron: 91 | Pulmonary delivery (general) | Elhissi and Taylor (2005) |

*API* active pharmaceutical ingredient, *DPPC* dipalmitoylphosphatidylcholine, *SPC* soy phosphatidylcholine, *VMD* volume median diameter, *%EE* percent entrapment efficiency, *FPF* fine particle fraction, *DMPC* dimyristoylphosphatidylcholine, *HSPC* hydrogenated soy phosphatidylcholine, *LMH* lactose monohydrate, *MCC* microcrystalline cellulose, *AmB* amphotericin B its unsuitability for the pulmonary administration of proliposome formulations, (Elhissi and Taylor 2005), agreeing with the previous impression of unsuitability of ultrasonic nebulizers for delivery of suspensions and viscous formulations (McCallion et al. 1995).

In an attempt to assess the efficiency of delivering nebulized proliposomes for medical indications, we have conducted another study to investigate the production and aerosolization of multilamellar and oligolamellar liposomes from ethanol-based SPC proliposome formulations carrying salbutamol sulphate (Elhissi et al. 2006b). The resulting liposomes were able to encapsulate up to 62% of the drug, a significant improvement compared to 1.23% encapsulation by conventionally prepared liposomes (Elhissi et al. 2006b). To determine their aerosol performance, the liposome formulations were aerosolized using an air-jet nebulizer (Pari LC Plus) and three vibrating-mesh nebulizers (Aeroneb Pro small mesh, Aeroneb Pro large mesh, Omron NE U22). Remarkably, the vibrating-mesh nebulizers produced larger aerosol droplet sizes with narrower size distributions compared to the air-jet nebulizers (Elhissi et al. 2006b). While the dispersion medium (sodium chloride or sucrose solution) had a minimal effect on the performance of the jet nebulizer, it significantly affected the performance of the three vibrating-mesh nebulizers. Compared with the air-jet nebulizer, vibrating-mesh devices caused less drug loss and vesicle size reduction during nebulization. Figure 3.13 provides a representative summary of the main findings reported in this study (Elhissi et al. 2006b).

In another exploration using the vibrating-mesh nebulizer, Aeroneb Go, ethanol-based proliposomes composed of SPC and cholesterol (1:1 mol ratio), were hydrated with NaCl (0.9%) or sucrose (9.25%) solutions to form liposomes encapsulating approximately 61% of the hydrophilic bronchodilator salbutamol sulfate (Elhissi et al. 2011a). Using the TSI,

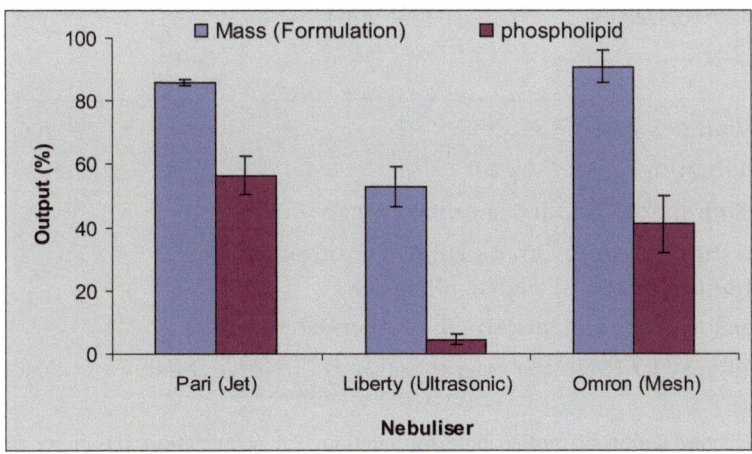

**Fig. 3.12** Effect of nebulizer design on total mass (formulation) and phospholipid outputs. Reproduced with permission from Elhissi and Taylor (2005). © Elsevier

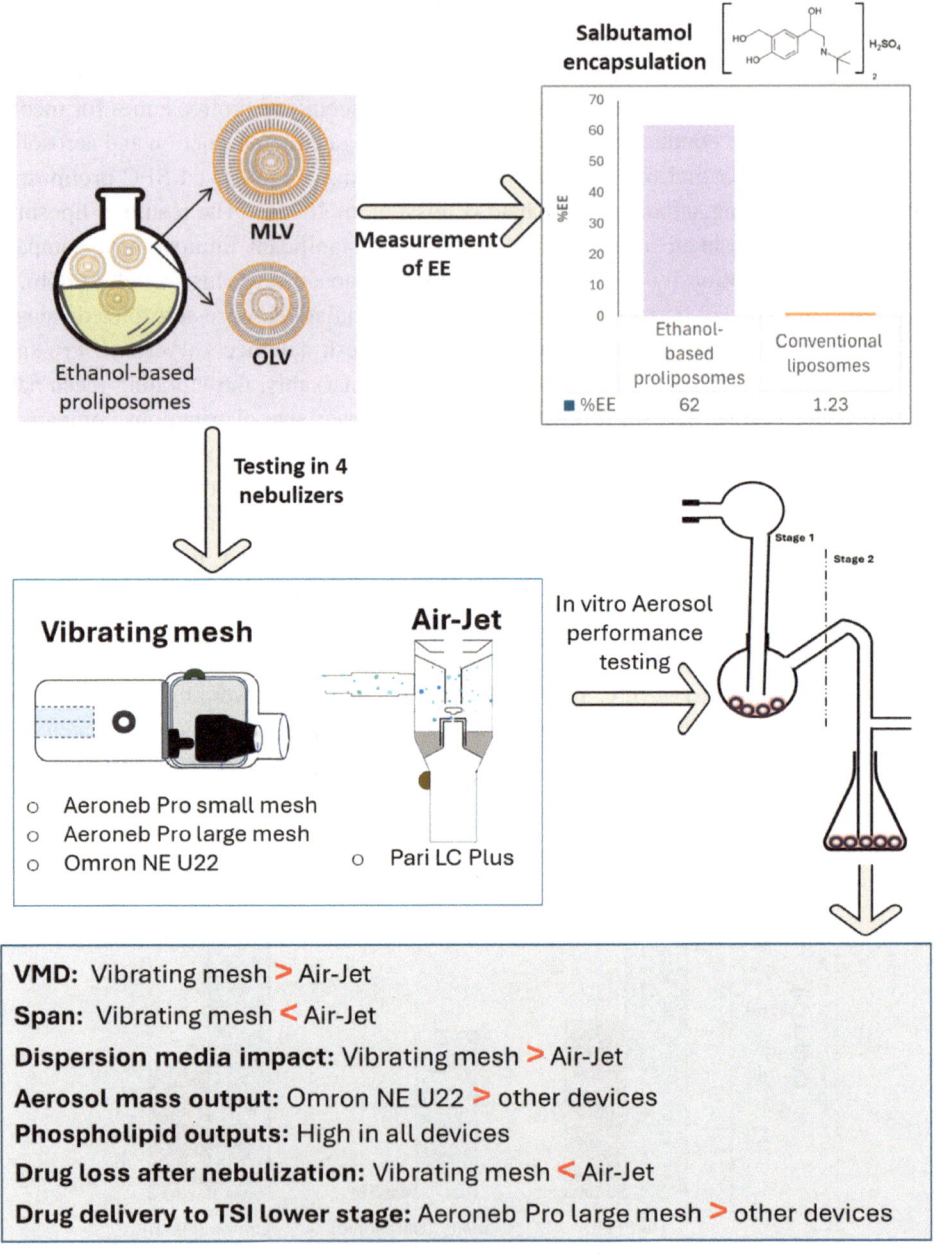

**Fig. 3.13** Representation of proliposome formulations for nebulization via air jet and vibrating mesh nebulizers and their pulmonary delivery study outcomes, as reported by Elhissi et al. (2006a, b)

both formulations had aerosol mass outputs exceeding the phospholipid output, suggesting some accumulation of large liposomes or lipid aggregates within the nebulizer during nebulization (Elhissi et al. 2011a). Notably, NaCl (0.9%) hydration medium, owing to its lower viscosity, produced smaller aerosol droplets and higher aerosol and phospholipid outputs than the sucrose solution (Elhissi et al. 2011a). Although FPF was relatively high (57.44%), liposome size reduction during nebulization led to significant drug losses (Elhissi et al. 2011a).

Furthermore, a high-sensitivity DSC was used to study BDP steroid entrapment in aerosolized liposomes produced from proliposome formulations. These proliposomes were hydrated in an air-jet nebulizer, forming liposome aerosols (Elhissi et al. 2011b). While BDP reduced aerosol and lipid outputs, analysis confirmed liposome formation and steroid incorporation into bilayers (Elhissi et al. 2011b). Analytical methods, including size analysis and TEM, confirmed the formation of liposomes within the nebulizer, followed by deaggregation and size reduction during nebulization to the impinger (Elhissi et al. 2011b). Thermal studies indicated the interaction between BDP and the liposome bilayers without causing phase separation, suggesting that the bilayers could entrap more than 5 mol% of this steroid (Elhissi et al. 2011b).

Furthermore, particulate-based proliposomes were prepared by coating sucrose particles with EPC, SPC, or SPC combined with a cholesterol (1:1) (Elhissi et al. 2012). The proliposomes generated multilamellar liposomes directly within the air-jet nebulizer, Pari LC Plus, upon hydration, eliminating the need for manual shaking prior to nebulization (Elhissi et al. 2012). All formulations delivered high aerosol and phospholipid outputs, with significant deposition in the lower stage of a TSI. However, liposomes containing SPC and cholesterol (1:1) exhibited greater rigidity, resulting in larger particle sizes and more lipid accumulation in the nebulizer (Elhissi et al. 2012).

Ethanol-based proliposomes encapsulating salbutamol sulfate have also been produced and explored for the generation and nebulization of liposomes via the Aeroneb Pro vibrating-mesh nebulizer (Elhissi et al. 2013). The produced liposomes showed improved performance, with FPF values of 57.85% compared with 45.81% for the conventional drug solution (Elhissi et al. 2013). While the aerosol VMDs were relatively similar (3.44 μm vs. 3.22 μm), the liposomal formulation achieved higher FPF due to reduced aerosol polydispersity, attributed to lower surface tension caused by the presence of phospholipids (Elhissi et al. 2013). Using the same production method, another study aimed to develop and evaluate chitosan-coated liposomes containing amphotericin B for delivery via an air-jet nebulizer (Albasarah et al. 2010). Multilamellar vesicles were generated via the use of ethanol-based proliposomes dispersed with sodium chloride (0.9%) solution or various concentrations of chitosan chloride (Albasarah et al. 2010). The results revealed that liposomes produced from proliposomes achieved the highest amphotericin B loading, with approximately 80% loading in the chitosan-coated formulations (Albasarah et al. 2010). Upon nebulization, approximately 60% of the amphotericin B from the liposomal formulations was deposited in the lower stage of the impinger (Fig. 3.14). Both chitosan-coated and uncoated liposomes demonstrated antifungal activity against *Candida albicans* and *Candida tropicalis*, comparable to that of amphotericin B deoxycholate micelles (MIC: 0.5 mg/mL) (Albasarah et al. 2010).

Moreover, Khan et al. developed novel paclitaxel (PTX)-loaded proliposome tablet formulations for pulmonary delivery via nebulization (Khan et al. 2020). Proliposome powders were prepared through sonication and spray drying using LMH, MCC, or starch as carriers and phospholipids such as SPC, HSPC, or DMPC (Khan et al. 2020). Among 27 formulations, three were selected based on superior flowability, vesicle size, and compressibility indices (Khan et al. 2020). The tablet formulations demonstrated optimal physical properties, including uniform weight, good crushing strength, and short disintegration time (~ 14.35 min) (Khan et al. 2020). Nebulization studies of the disintegrated tablets revealed that ultrasonic nebulizers provided better performance, with shorter nebulization times (8.75 min) and higher output rates (421.06 mg/min) than vibrating-mesh nebulizers (Khan et al. 2020). Tablets consisting from LMH carrier at a 1:25 lipid-to-carrier have demonstrated selective toxicity against cancer cells in vitro, while being less toxic to normal cells (Fig. 3.15) (Khan et al. 2020).

In another study, proliposome powder and tablet formulations were developed for delivery of BDP via the air-jet nebulizer Pari-LC Sprint (Khan et al. 2021). The proliposome powders were prepared with sorbitol or mannitol as carriers at 1:10 and 1:15 w/w lipid–carrier ratios and then compressed into tablets. SEM images of the six manufactured proliposome formulations and TEM images of the generated liposomes from the two formulations are shown in Fig. 3.16. Post-nebulization, the liposome size (7.79–9.73 μm) was larger than that of freshly prepared liposomes (5.38–5.85 μm), likely due

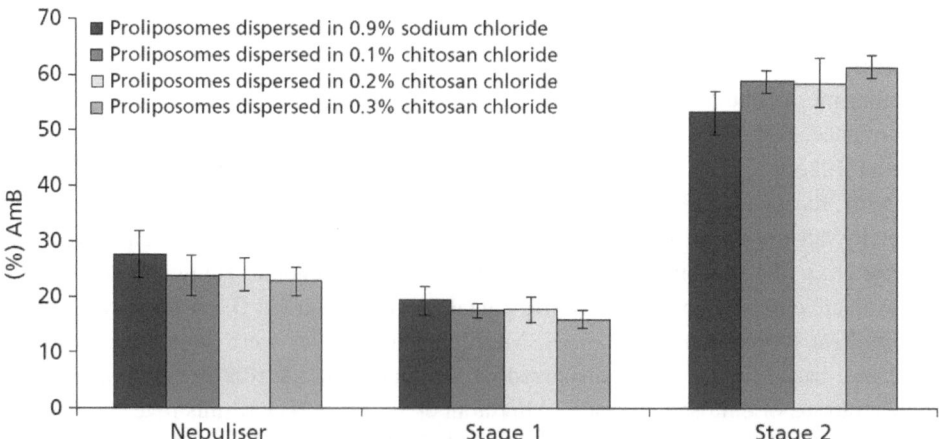

**Fig. 3.14** Deposition of amphotericin B in TSI and remaining in the nebulizer following nebulization of proliposomes dispersed in 0.9% sodium chloride and different concentrations of chitosan chloride. Reproduced with permission from Albasarah et al. (2010). © Oxford University Press

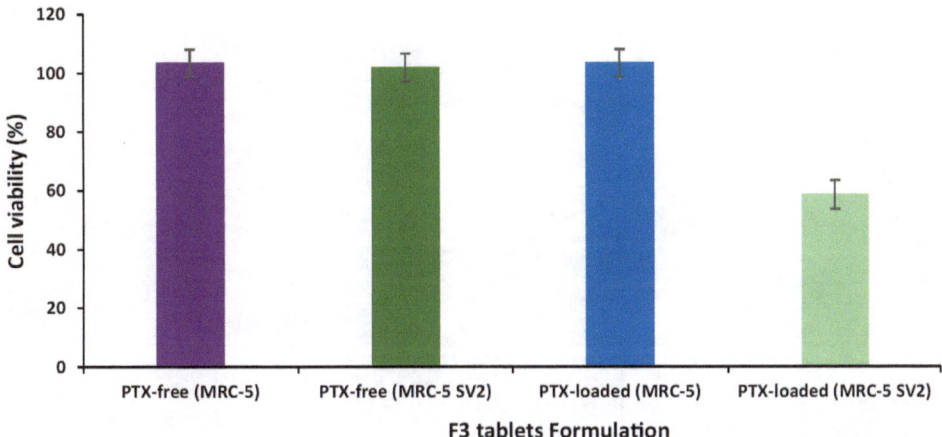

**Fig. 3.15** Viability of the MRC-5 and MRC-5 SV2 cell lines tested with transfersomes generated from a proliposome tablet formulation (LMH carrier at a 1:25 lipid-to-carrier ratio) via both PTX-free and PTX-loaded formulations in black, flat-bottomed 96-well plates. Reproduced from Khan et al. (2020), under the Creative Commons Attribution License (CC BY)

to aggregation in the nebulizer reservoir (Khan et al. 2021). Both powders and tablets achieved over 80% mass output, with tablets delivering a greater proportion of the drug (~ 50%) to the lower stage of the TSI (Khan et al. 2021). As depicted in Fig. 3.17, over 8 h, liposomes generated from mannitol-based proliposome powders and tablets presented the highest drug release rates in both the 1:10 and the 1:15 w/w formulations (Khan et al. 2021). In contrast, sorbitol-based proliposome powders and tablets exhibited significantly slower BDP release profiles ($p < 0.05$) (Khan et al. 2021). Together, these findings suggest that proliposome tablets can effectively generate inhalable aerosols while ensuring sustained BDP release.

In another study, the antiviral drug favipiravir was encapsulated in a proliposomal powder using the film deposition on carrier method (Shaik et al. 2022). The formulation exhibited good flow properties, with liposome vesicles forming within 2–3 min following hydration (Shaik et al. 2022). In vitro nebulization studies using an air-jet nebulizer and a TSI demonstrated nebulization times of 12.95–14.3 min, aerosol mass outputs of 44.31–54.28%, and respirable fractions of 8.14–12.31% (Shaik et al. 2022). While these findings indicate the applicability of proliposomal formulations for different pulmonary indications, they warrant further exploration to enhance the nebulization outcomes by optimizing the formulation and delivery method.

In summary, the use of proliposomes in nebulization for pulmonary drug delivery offers significant advantages, such as improved drug entrapment, prolonged therapeutic effects, and reduced systemic adverse effects. However, the effective delivery of proliposome-generated liposomes through nebulizers presents unique challenges. These challenges

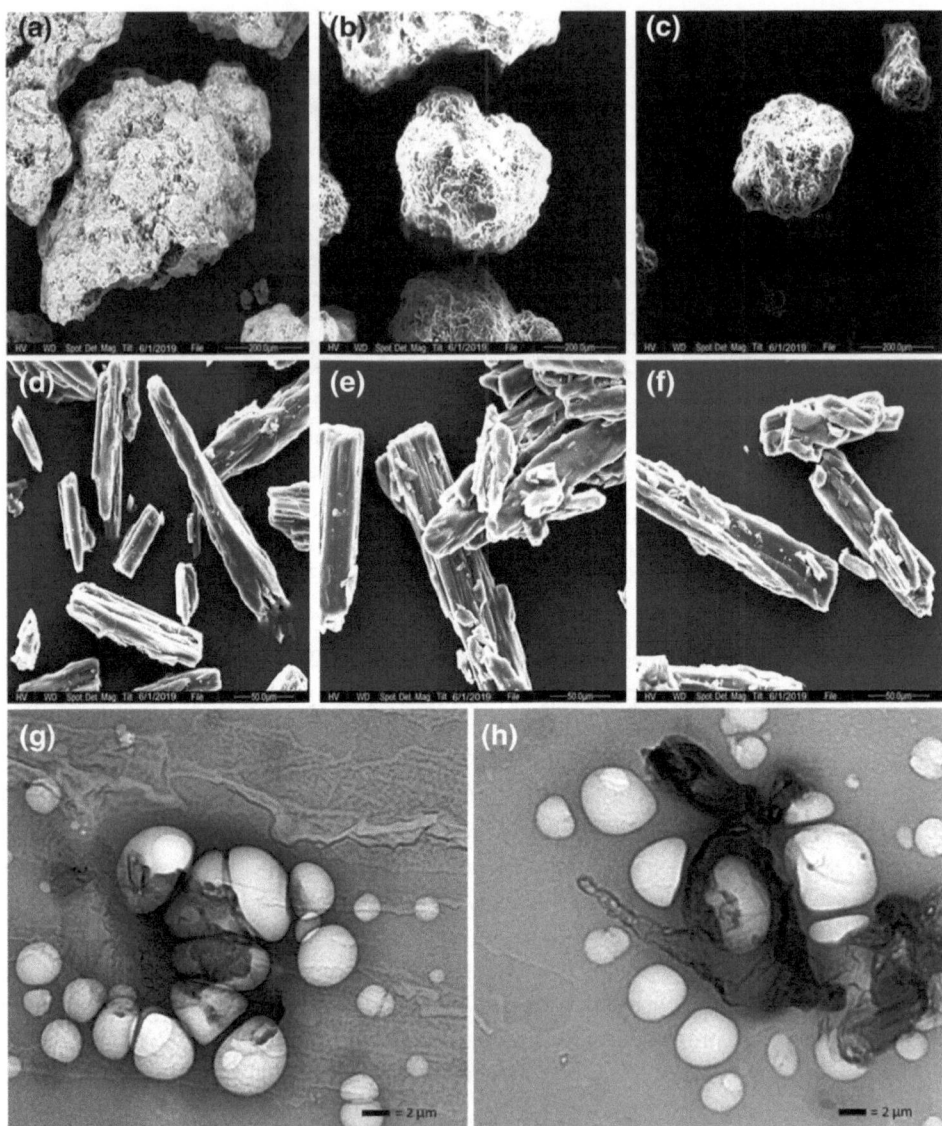

**Fig. 3.16** SEM images of coarse carbohydrate carrier and proliposome powders prepared with different w/w lipid phase-to-carrier ratios: **a** coarse sorbitol, **b** sorbitol-based proliposomes with 1:10 w/w, **c** sorbitol-based proliposomes with 1:15 w/w, **d** coarse mannitol, **e** mannitol-based proliposomes with 1:10 w/w, and **f** mannitol-based proliposomes with 1:15 w/w. TEM images of liposomes generated from **g** sorbitol-based proliposome powder at a 1:10 w/w ratio and **h** sorbitol-based proliposome tablets at a 1:10 w/w ratio. Reproduced from Khan et al. (2021), under the Creative Commons Attribution License (CC BY)

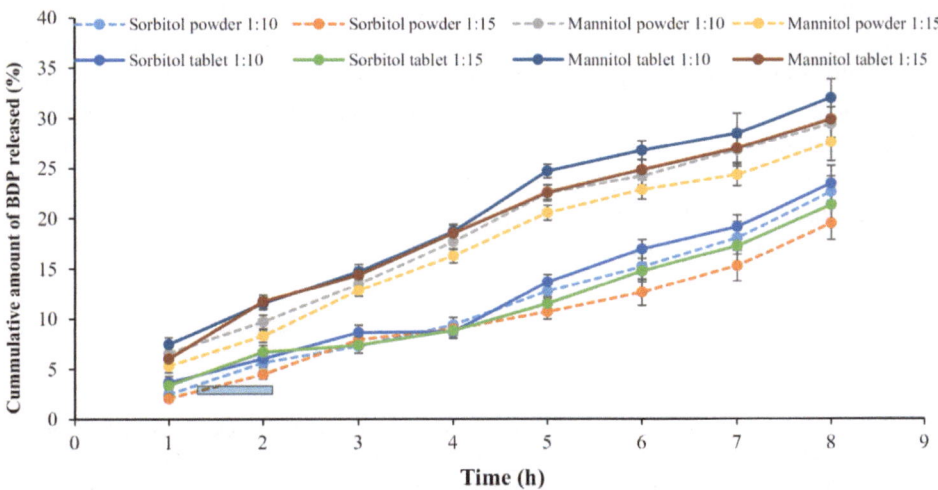

**Fig. 3.17** In vitro release of BDP from liposomes generated from sorbitol or mannitol-based pro-liposome powder and tablet formulations for both 1:10 and 1:15 w/w lipid-to-carrier ratios. Reproduced from Khan et al. (2021), under the Creative Commons Attribution License (CC BY)

arise from device-specific performance issues and formulation design, all of which may impact the efficiency and practicality of formulation delivery. Understanding these limitations is essential for optimizing both formulations and nebulizer designs to achieve reliable and effective therapeutic outcomes. For example, the nebulization of proliposomes through various devices revealed notable issues related to drug entrapment and stability. Device-specific performance challenges may exacerbate these issues, as noted by different performance outcomes based on the nebulizer type and function. Formulation-specific limitations also play a critical role in the performance of proliposome nebulization. Output inefficiencies may also limit the practicality of proliposome-based nebulization. For example, issues such as phospholipid accumulation in the nebulizer reservoir and inconsistent aerosol mass and phospholipid outputs may reduce therapeutic efficacy.

## 3.4    Conclusion

Proliposomes offer several benefits for pulmonary drug delivery because of their ability to transform into liposomes upon contact with pulmonary fluids. However, the efficiency of proliposomes is influenced by the formulation approach and the type of inhalation device used for delivery. Different devices, including pMDIs, DPIs, and nebulizers, have been explored for the delivery of proliposomes for pulmonary delivery. While pMDIs were initially explored for proliposome delivery, the phase-out of CFC propellants and the incompatibility of alternative HFA propellants with phospholipids shifted the attention toward alternative devices, including DPIs and medical nebulizers. DPIs have been successfully used to deliver proliposome formulations for various indications. Further advancements in formulation techniques and tailored excipient combinations can broaden their applicability and clinical impact. Nebulizers have also been explored for delivering proliposomes, but factors such as device-specific performance, formulation design, and potential fragmentation of liposomes during nebulization need to be addressed for optimal therapeutic outcomes.

# References

Adel, I. M., ElMeligy, M. F., Abdelrahim, M. E., Maged, A., Abdelkhalek, A. A., Abdelmoteleb, A. M., & Elkasabgy, N. A. (2021). Design and Characterization of Spray-Dried Proliposomes for the Pulmonary Delivery of Curcumin. *International Journal of Nanomedicine, Volume 16*, 2667–2687. https://doi.org/10.2147/IJN.S306831

Aekwattanaphol, N., Das, S. C., Khadka, P., Nakpheng, T., Ali Khumaini Mudhar Bintang, M., & Srichana, T. (2024). Development of a proliposomal pretomanid dry powder inhaler as a novel alternative approach for combating pulmonary tuberculosis. *International Journal of Pharmaceutics, 664*, 124608. https://doi.org/10.1016/j.ijpharm.2024.124608

Albasarah, Y. Y., Somavarapu, S., Stapleton, P., & Taylor, K. M. G. (2010). Chitosan-coated antifungal formulations for nebulisation. *Journal of Pharmacy and Pharmacology, 62*(7), 821–828. https://doi.org/10.1211/jpp.62.07.0002

Bibi, S., Kaur, R., Henriksen-Lacey, M., McNeil, S. E., Wilkhu, J., Lattmann, E., Christensen, D., Mohammed, A. R., & Perrie, Y. (2011). Microscopy imaging of liposomes: From coverslips to environmental SEM. *International Journal of Pharmaceutics, 417*(1–2), 138–150. https://doi.org/10.1016/j.ijpharm.2010.12.021

Bonechi, C., Tamasi, G., Donati, A., Leone, G., Consumi, M., Cangeloni, L., Volpi, V., Magnani, A., Cappelli, A., & Rossi, C. (2021). Physicochemical Characterization of Hyaluronic Acid and Chitosan Liposome Coatings. *Applied Sciences, 11*(24), 12071. https://doi.org/10.3390/app112412071

Choudhary, M., Chaurawal, N., Barkat, Md. A., & Raza, K. (2022). Proliposome-Based Nanostrategies: Challenges and Development as Drug Delivery Systems. *AAPS PharmSciTech, 23*(8), 293. https://doi.org/10.1208/s12249-022-02443-1

Dallal Bashi, Y. H., Ali, A., Al Ayoub, Y., Assi, K. H., Mairs, R., McCarthy, H. O., Tunney, M. M., & Kett, V. L. (2024). Inhaled dry powder liposomal azithromycin for treatment of chronic lower respiratory tract infection. *International Journal of Pharmaceutics, 653*, 123841. https://doi.org/10.1016/j.ijpharm.2024.123841

Danaei, M., Dehghankhold, M., Ataei, S., Hasanzadeh Davarani, F., Javanmard, R., Dokhani, A., Khorasani, S., & Mozafari, M. R. (2018). Impact of Particle Size and Polydispersity Index on the Clinical Applications of Lipidic Nanocarrier Systems. *Pharmaceutics, 10*(2), 57. https://doi.org/10.3390/pharmaceutics10020057

Dhiman, N., Sarvaiya, J., & Mohindroo, P. (2022). A drift on liposomes to proliposomes: recent advances and promising approaches. *Journal of Liposome Research, 32*(4), 317–331. https://doi.org/10.1080/08982104.2021.2019762

Elhissi, A. (2017). Liposomes for Pulmonary Drug Delivery: The Role of Formulation and Inhalation Device Design. *Current Pharmaceutical Design, 23*(3), 362–372. https://doi.org/10.2174/138161 2823666161116114732

Elhissi, A. M. A., & Taylor, K. M. G. (2005). Delivery of liposomes generated from proliposomes using air-jet, ultrasonic, and vibrating-mesh nebulisers. *Journal of Drug Delivery Science and Technology, 15*(4), 261–265. https://doi.org/10.1016/S1773-2247(05)50047-9

Elhissi, A., Oneill, M., Roberts, S., & Taylor, K. (2006a). A calorimetric study of dimyristoylphosphatidylcholine phase transitions and steroid–liposome interactions for liposomes prepared by thin film and proliposome methods. *International Journal of Pharmaceutics, 320*(1–2), 124–130. https://doi.org/10.1016/j.ijpharm.2006.04.015

Elhissi, A., Karnam, K., Danesh-Azari, M.-R., Gill, H., & Taylor, K. (2006b). Formulations generated from ethanol-based proliposomes for delivery via medical nebulizers. *Journal of Pharmacy and Pharmacology, 58*(7), 887–894. https://doi.org/10.1211/jpp.58.7.0002

Elhissi, A., Gill, H., Ahmed, W., & Taylor, K. (2011a). Vibrating-mesh nebulization of liposomes generated using an ethanol-based proliposome technology. *Journal of Liposome Research, 21*(2), 173–180. https://doi.org/10.3109/08982104.2010.505574

Elhissi, A. M. A., O'Neill, M., Ahmed, W., & Taylor, K. M. G. (2011b). High-sensitivity differential scanning calorimetry for measurement of steroid entrapment in nebulised liposomes generated from proliposomes. *Micro & Nano Letters, 6*(8), 694–697. https://doi.org/10.1049/mnl.2011. 0086

Elhissi, A., Ahmed, W., & Taylor, K. M. G. (2012). Laser Diffraction and Electron Microscopy Studies on Inhalable Liposomes Generated from Particulate-Based Proliposomes Within a Medical Nebulizer. *Journal of Nanoscience and Nanotechnology, 12*(8), 6693–6699. https://doi.org/10. 1166/jnn.2012.4566

Elhissi, Brar Jasmeet, Najlah Mohammad, Roberts Simon, Faheem Ahmed, & Taylor Kevin. (2013). An Ethanol-Based Proliposome Technology for Enhanced Delivery and Improved "Respirability" of Antiasthma Aerosols Generated Using a Micropump Vibrating-Mesh Nebulizer. *Journal of Pharmaceutical Technology, Research and Management, 1*(2), 171–180. https://doi.org/10.15415/ jptrm.2013.12010

Farr, S. J., Kellaway, I. W., & Carman-Meakin, B. (1987). Assessing the potential of aerosol-generated liposomes from pressurised pack formulations. *Journal of Controlled Release, 5*(2), 119–127. https://doi.org/10.1016/0168-3659(87)90003-4

Finlay, W. H., & Darquenne, C. (2020). Particle Size Distributions. *Journal of Aerosol Medicine and Pulmonary Drug Delivery, 33*(4), 178–180. https://doi.org/10.1089/jamp.2020.29028.whf

Gomez, A. I., Acosta, M. F., Muralidharan, P., Yuan, J. X.-J., Black, S. M., Hayes, D., & Mansour, H. M. (2020). Advanced spray dried proliposomes of amphotericin B lung surfactant-mimic phospholipid microparticles/nanoparticles as dry powder inhalers for targeted pulmonary drug delivery. *Pulmonary Pharmacology & Therapeutics, 64*, 101975. https://doi.org/10.1016/j.pupt. 2020.101975

Jin, X., Yu, H., Zhang, Z., Cui, T., Wu, Q., Liu, X., Gao, J., Zhao, X., Shi, J., Qu, G., & Jiang, G. (2021). Surface charge-dependent mitochondrial response to similar intracellular nanoparticle contents at sublethal dosages. *Particle and Fibre Toxicology, 18*(1), 36. https://doi.org/10.1186/ s12989-021-00429-8

Khan, I., Yousaf, S., Subramanian, S., Korale, O., Alhnan, M. A., Ahmed, W., Taylor, K. M. G., & Elhissi, A. (2015). Proliposome powders prepared using a slurry method for the generation of beclometasone dipropionate liposomes. *International Journal of Pharmaceutics, 496*(2), 342–350. https://doi.org/10.1016/j.ijpharm.2015.10.002

Khan, I., Yousaf, S., Subramanian, S., Albed Alhnan, M., Ahmed, W., & Elhissi, A. (2018). Prolipo-some tablets manufactured using a slurry-driven lipid-enriched powders: Development, charac-terization and stability evaluation. *International Journal of Pharmaceutics*, *538*(1–2), 250–262. https://doi.org/10.1016/j.ijpharm.2017.12.049

Khan, I., Lau, K., Bnyan, R., Houacine, C., Roberts, M., Isreb, A., Elhissi, A., & Yousaf, S. (2020). A Facile and Novel Approach to Manufacture Paclitaxel-Loaded Proliposome Tablet Formulations of Micro or Nano Vesicles for Nebulization. *Pharmaceutical Research*, *37*(6), 116. https://doi.org/10.1007/s11095-020-02840-w

Khan, I., Yousaf, S., Najlah, M., Ahmed, W., & Elhissi, A. (2021). Proliposome powder or tablets for generating inhalable liposomes using a medical nebulizer. *Journal of Pharmaceutical Inves-tigation*, *51*(1), 61–73. https://doi.org/10.1007/s40005-020-00495-8

Khan, I., Al-Hasani, A., Khan, M. H., Khan, A. N., -Alam, F., Sadozai, S. K., Elhissi, A., Khan, J., & Yousaf, S. (2023). Impact of dispersion media and carrier type on spray-dried proliposome pow-der formulations loaded with beclomethasone dipropionate for their pulmonary drug delivery via a next generation impactor. *PLOS ONE*, *18*(3), e0281860. https://doi.org/10.1371/journal.pone.0281860

Knap, K., Kwiecień, K., Reczyńska-Kolman, K., & Pamuła, E. (2023). Inhalable microparticles as drug delivery systems to the lungs in a dry powder formulations. *Regenerative Biomaterials*, *10*. https://doi.org/10.1093/rb/rbac099

Lombardo, D., & Kiselev, M. A. (2022). Methods of Liposomes Preparation: Formation and Control Factors of Versatile Nanocarriers for Biomedical and Nanomedicine Application. *Pharmaceutics*, *14*(3), 543. https://doi.org/10.3390/pharmaceutics14030543

McCallion, O. N., Taylor, K. M., Thomas, M., Taylor, A. J. (1995). Nebulization of fluids of different physicochemical properties with air-jet and ultrasonic nebulizers. Pharm Res, 12(11):1682–1688. https://doi.org/10.1023/a:1016205520044

Mišík, O., Kejíková, J., Cejpek, O., Malý, M., Jugl, A., Bělka, M., Mravec, F., & Lízal, F. (2024). Nebulization and *In Vitro* Upper Airway Deposition of Liposomal Carrier Systems. *Molecular Pharmaceutics*, *21*(4), 1848–1860. https://doi.org/10.1021/acs.molpharmaceut.3c01146

Németh, Z., Csóka, I., Semnani Jazani, R., Sipos, B., Haspel, H., Kozma, G., Kónya, Z., & Dobó, D. G. (2022). Quality by Design-Driven Zeta Potential Optimisation Study of Liposomes with Charge Imparting Membrane Additives. *Pharmaceutics*, *14*(9), 1798. https://doi.org/10.3390/pharmaceutics14091798

Omer, H. K., Hussein, N. R., Ferraz, A., Najlah, M., Ahmed, W., Taylor, K. M. G., & Elhissi, A. M. A. (2018). Spray-Dried Proliposome Microparticles for High-Performance Aerosol Delivery Using a Monodose Powder Inhaler. *AAPS PharmSciTech*, *19*(5), 2434–2448. https://doi.org/10.1208/s12249-018-1058-4

Öztürk, K., Kaplan, M., & Çalış, S. (2024). Effects of nanoparticle size, shape, and zeta potential on drug delivery. *International Journal of Pharmaceutics*, *666*, 124799. https://doi.org/10.1016/j.ijpharm.2024.124799

Parhizkar, E., Sadeghinia, D., Hamishehkar, H., Yaqoubi, S., Nokhodchi, A., & Alipour, S. (2021). Carrier Effect in Development of Rifampin Loaded Proliposome for Pulmonary Delivery: A Quality by Design Study. *Advanced Pharmaceutical Bulletin*. https://doi.org/10.34172/apb.2022.032

Patil-Gadhe, A., & Pokharkar, V. (2013). Single step spray drying method to develop proliposomes for inhalation: A systematic study based on quality by design approach. *Pulmonary Pharmacol-ogy & Therapeutics*, *27*(2), 197–207. https://doi.org/10.1016/j.pupt.2013.07.006

Peleg-Shulman, T., Gibson, D., Cohen, R., Abra, R., & Barenholz, Y. (2001). Characterization of sterically stabilized cisplatin liposomes by nuclear magnetic resonance. *Biochimica et Biophysica*

*Acta (BBA) - Biomembranes, 1510*(1–2), 278–291. https://doi.org/10.1016/S0005-2736(00)003 59-X

Pochapski, D. J., Carvalho dos Santos, C., Leite, G. W., Pulcinelli, S. H., & Santilli, C. V. (2021). Zeta Potential and Colloidal Stability Predictions for Inorganic Nanoparticle Dispersions: Effects of Experimental Conditions and Electrokinetic Models on the Interpretation of Results. *Langmuir, 37*(45), 13379–13389. https://doi.org/10.1021/acs.langmuir.1c02056

Pokharkar, V., Patil-Gadhe, A., Kyadarkunte, A., Pereira, M., Jejurikar, G., Patole, M., & Risbud, A. (2014). Rifapentine-proliposomes for inhalation: In vitro and In vivo toxicity. *Toxicology International, 21*(3), 275. https://doi.org/10.4103/0971-6580.155361

Rojanarat, W., Changsan, N., Tawithong, E., Pinsuwan, S., Chan, H.-K., & Srichana, T. (2011). Isoniazid Proliposome Powders for Inhalation—Preparation, Characterization and Cell Culture Studies. *International Journal of Molecular Sciences, 12*(7), 4414–4434. https://doi.org/10.3390/ijms12074414

Rojanarat, W., Nakpheng, T., Thawithong, E., Yanyium, N., & Srichana, T. (2012a). Inhaled pyrazinamide proliposome for targeting alveolar macrophages. *Drug Delivery, 19*(7), 334–345. https://doi.org/10.3109/10717544.2012.721144

Rojanarat, W., Nakpheng, T., Thawithong, E., Yanyium, N., & Srichana, T. (2012b). Levofloxacin-Proliposomes: Opportunities for Use in Lung Tuberculosis. *Pharmaceutics, 4*(3), 385–412. https://doi.org/10.3390/pharmaceutics4030385

Rudokas, M., Najlah, M., Alhnan, M. A., & Elhissi, A. (2016). Liposome Delivery Systems for Inhalation: A Critical Review Highlighting Formulation Issues and Anticancer Applications. *Medical Principles and Practice, 25*(Suppl. 2), 60–72. https://doi.org/10.1159/000445116

Ruiz, V. H., Encinas-Basurto, D., Ortega-Alarcon, N., Eedara, B. B., Fineman, J. R., Black, S. M., & Mansour, H. M. (2024). Inhalable Advanced Co-Spray Dried Microparticles/Nanoparticles of a Novel RhoA/Rho Kinase Inhibitor with Lung Surfactant Biomimetic Phospholipids for Targeted Lung Delivery. *ACS Pharmacology & Translational Science, 7*(10), 3241–3254. https://doi.org/10.1021/acsptsci.4c00432

Serrano-Lotina, A., Portela, R., Baeza, P., Alcolea-Rodriguez, V., Villarroel, M., & Ávila, P. (2023). Zeta potential as a tool for functional materials development. *Catalysis Today, 423*, 113862. https://doi.org/10.1016/j.cattod.2022.08.004

Shaik, N. B., Lakshmi, P. K., & Basava Rao, V. V. (2022). Formulation and Evaluation of Favipiravir Proliposomal Powder for Pulmonary Delivery by Nebulization. *International Journal of Pharmaceutical Research and Allied Sciences, 11*(2), 36–44. https://doi.org/10.51847/4McfhPccXs

Shao, X., Wei, X., Song, X., Hao, L., Cai, X., Zhang, Z., Peng, Q., & Lin, Y. (2015). Independent effect of polymeric nanoparticle zeta potential/surface charge, on their cytotoxicity and affinity to cells. *Cell Proliferation, 48*(4), 465–474. https://doi.org/10.1111/cpr.12192

Shi, Y., & Li, X. (2023). *High-Performance Liquid Chromatography Coupled with Tandem Mass Spectrometry Method for the Identification and Quantification of Lipids in Liposomes* (pp. 227–239). https://doi.org/10.1007/978-1-0716-2954-3_20

Singh, P., Bodycomb, J., Travers, B., Tatarkiewicz, K., Travers, S., Matyas, G. R., & Beck, Z. (2019). Particle size analyses of polydisperse liposome formulations with a novel multispectral advanced nanoparticle tracking technology. *International Journal of Pharmaceutics, 566*, 680–686. https://doi.org/10.1016/j.ijpharm.2019.06.013

Sommerville, M. L., Johnson, C. S., Cain, J. B., Rypacek, F., & Hickey, A. J. (2002). Lecithin Microemulsions in Dimethyl Ether and Propane for the Generation of Pharmaceutical Aerosols Containing Polar Solutes. In *Pharmaceutical Development and Technology* (Vol. 7, Issue 3). www.dekker.com

Srichana, T., Eze, F. N., & Thawithong, E. (2022). A facile one-step jet-milling approach for the preparation of proliposomal dry powder for inhalation as effective delivery system for anti-TB therapeutics. *Drug Development and Industrial Pharmacy, 48*(10), 528–538. https://doi.org/10.1080/03639045.2022.2135101

Srichana, T., Thawithong, E., Nakpheng, T., & Paul, P. K. (2023). Flow cytometric analysis, confocal laser scanning microscopic, and holotomographic imaging demonstrate potentials of levofloxacin dry powder aerosols for TB treatment. *Journal of Drug Delivery Science and Technology, 84*, 104464. https://doi.org/10.1016/j.jddst.2023.104464

Stewart, J. C. M. (1980). Colorimetric determination of phospholipids with ammonium ferrothiocyanate. *Analytical Biochemistry, 104*(1), 10–14. https://doi.org/10.1016/0003-2697(80)90269-9

Tarara, T. E., Hartman, M. S., Gill, H., Kennedy, A. A., & Weers, J. G. (2004). Characterization of Suspension-Based Metered Dose Inhaler Formulations Composed of Spray-Dried Budesonide Microcrystals Dispersed in HFA-134a. *Pharmaceutical Research, 21*(9), 1607–1614. https://doi.org/10.1023/B:PHAM.0000041455.13980.f1

US Pharmacopeia. (2015). *Bulk Density and Tapped Density of Powders*

Vyas, S. P., & Sakthivel, T. (1994). Pressurized pack-based liposomes for pulmonary targeting of isoprenaline-development and characterization. In *J. MICROENCAPSULATION* (Vol. 1, Issue 1)

Vyas, S. P., Kannan, M. E., Jain, S., Mishra, V., & Singh, P. (2004). Design of liposomal aerosols for improved delivery of rifampicin to alveolar macrophages. *International Journal of Pharmaceutics, 269*(1), 37–49. https://doi.org/10.1016/j.ijpharm.2003.08.017

Vyas, S. P., Quraishi, S., Gupta, S., & Jaganathan, K. S. (2005). Aerosolized liposome-based delivery of amphotericin B to alveolar macrophages. *International Journal of Pharmaceutics, 296*(1–2), 12–25. https://doi.org/10.1016/j.ijpharm.2005.02.003

Wang, B., Wang, L., Yang, Q., Zhang, Y., Qinglai, T., Yang, X., Xiao, Z., Lei, L., & Li, S. (2024a). Pulmonary inhalation for disease treatment: Basic research and clinical translations. *Materials Today Bio, 25*, 100966. https://doi.org/10.1016/j.mtbio.2024.100966

Wang, J., Guo, Y., Lu, W., Liu, X., Zhang, J., Sun, J., & Chai, G. (2024b). Dry powder inhalation containing muco-inert ciprofloxacin and colistin co-loaded liposomes for pulmonary *P. aeruginosa* biofilm eradication. *International Journal of Pharmaceutics, 658*, 124208. https://doi.org/10.1016/j.ijpharm.2024.124208

# Therapeutic Applications of Proliposomes in Pulmonary Drug Delivery

**4**

## Abstract

In this chapter, the published articles pertinent to therapeutic applications of inhalable proliposome drug delivery systems, or liposomes generated from proliposomes were reviewed and summarized. It was found that these applications focused highly on pulmonary infectious diseases such as tuberculosis, pseudomonas infections and fungal infections. Incorporation of a wide range of hydrophilic (e.g. aminoglycosides) and hydrophobic (Amphotericin B) antibiotic drugs into proliposome formulations was discussed, and the role of formulation including carrier type, lipid composition and inclusion of bioadhesives was evaluated. The potential of proliposomes in the treatment of pulmonary infections is derived from the established success and FDA approval of the non-proliposomal liposome formulation of amikacin (Arikayce®) designed for the treatment of *Mycobacterium avium* complex (MAC) pulmonary infection. Importantly, the application of proliposome drug delivery systems for the treatment of other diseases was also highlighted, such as asthma, cancer, and in gene therapy. The promising findings achieved with proliposomes including in vitro investigations and in vivo studies using experimental animals, suggest possible clinical investigations to take place in the future.

## 4.1    Introduction

Proliposomes have emerged as versatile and promising drug delivery systems within the field of nanotechnology, with great promise for addressing some of the critical limitations associated with traditional liposome formulations (Choudhary et al. 2022; Dhiman et al. 2022; Elhissi 2017). One formulation approach of proliposomes is the preparation of free-flowing granules/powders of carbohydrate incorporating phospholipid, that generate liposomes upon contact with an aqueous phase above the phase transition temperature of the lipid (Choudhary et al. 2022; Dhiman et al. 2022; Elhissi 2017). This unique feature enhances their stability compared with that of liquid liposomal dispersions, which are prone to instability manifestations (Dhiman et al. 2022; Huang et al. 2024). By overcoming these stability challenges, proliposomes offer a robust solution as pulmonary drug delivery system with extended shelf life and maintained therapeutic effects.

Liposomes and proliposomes can deliver medications directly to the respiratory tract, making them highly effective for treating localized lung conditions such as asthma, lung cancer, cystic fibrosis, and pulmonary infections. In addition, these formulations have potential for systemic drug delivery through the lungs, offering controlled release and sustained therapeutic effects that reduce the frequency of dosing and minimize systemic adverse effects.

Remarkably, some liposomal formulations have already been approved for the management of a range of pulmonary conditions, such as liposomal amikacin (Arikayce®) for the treatment of *Mycobacterium avium* complex (MAC) lung disease (Ferreira et al. 2021). The delivery of these formulations not only improves drug stability and bioavailability but also enables targeted therapy and reduces toxicity. In particular, the enhanced stability and ease of storage make proliposomes predominantly suitable for developing controlled-release pulmonary formulations.

Current research continues to highlight the potential of proliposomes in advancing pulmonary drug delivery. Owing to their superior stability and biocompatibility, their ability to incorporate both hydrophilic and hydrophobic drugs represents crucial innovations in the treatment of various respiratory and systemic conditions via the pulmonary route. In this chapter, we provide a review of the applications of proliposomes as a formulation approach for generating liposomes and as a drug delivery system by itself, for the treatment of various diseases via inhalation. Particularly, proliposomes have been developed to target respiratory tract infections (i.e., tuberculosis, *Pseudomonas aeruginosa* infections, and fungal infections), asthma, lung cancer, and gene therapy, as illustrated in Fig. 4.1. The review in this chapter includes in vitro studies using tissue culture and in vivo investigations that employed experimental animals, with a highlight on the potential for future clinical applications of proliposomes and liposomes generated from proliposomes.

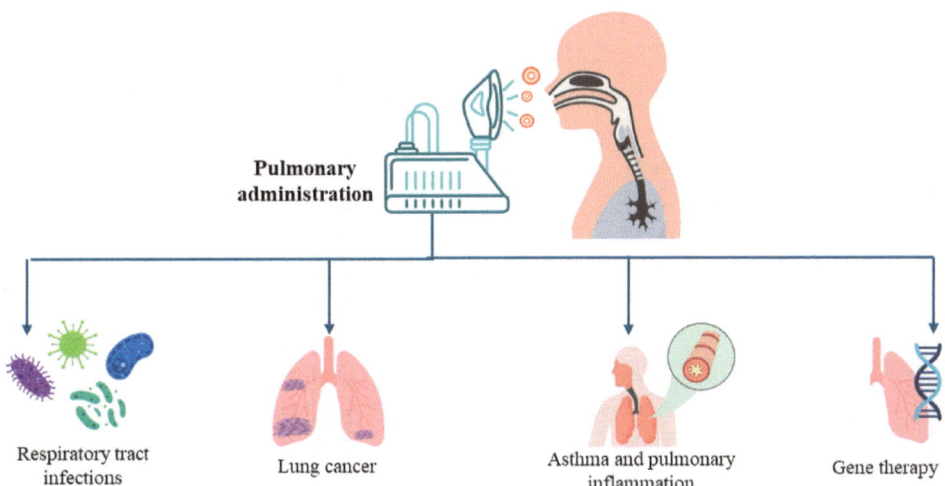

**Fig. 4.1** Current reported therapeutic applications of proliposomes in the pulmonary delivery of medications

## 4.2    Therapeutic Applications

### 4.2.1    Respiratory Tract Infections

#### 4.2.1.1    Tuberculosis

Tuberculosis (TB) is a contagious infection caused by *Mycobacterium tuberculosis* that spreads among humans through the respiratory system (Alsayed and Gunosewoyo 2023; Smith 2003). The main challenge in treating TB is the ability of *M. tuberculosis*, the main pathogen responsible for TB, to thrive within macrophages and increase the rate of antimicrobial resistance (Alsayed and Gunosewoyo 2023; Smith 2003). Because of their promising characteristics, proliposomes are appropriate for the targeted treatment of TB due to their improved stability and potential in overcoming the current limitations of available therapeutics. The main studies that reported the successful design and use of proliposomal formulations for potential pulmonary administration and management of mycobacterial infections are summarized in Table 4.1. Additionally, the main findings of these studies are illustrated in Fig. 4.2.

One promising approach for tackling TB involves the use of proliposomes for the encapsulation of the first-line TB medication rifampicin and its analogues. For instance, particulate-based spray-dried proliposomes loaded with rifampicin were prepared by Parhizkar et al. (2021). The proliposomal preparation displayed favorable in vitro aerosolization properties upon hydration to form liposomes with more efficient and targeted treatment (Parhizkar et al. 2021). Additionally, rifapentine, a rifampicin analogue,

**Table 4.1** Summary of studies on the pulmonary delivery of proliposomal formulations for the treatment of different respiratory tract infections

| Condition | Medication | Formulation | Main findings | Author and year |
|---|---|---|---|---|
| Tuberculosis | Pretomanid | Spray-dried proliposomes with L-leucine and trehalose | Pretomanid proliposomes showed high drug content, suitable aerosolization properties, and upon hydration a higher antimycobacterial activity than pretomanid alone. It also demonstrated to be safe for broncho-epithelial cells | Aekwattanaphol et al. (2024) |
| Tuberculosis | Delamanid | Proliposomes with human serum albumin (HSA) | Upon hydration of proliposomes, delamanid showed strong, spontaneous binding to HSA, potentially affecting stability and bioavailability of the drug when formulated in proliposomes | Tongkanarak et al. (2024) |
| Tuberculosis | Isoniazid and rifampicin | Jet-milled proliposomes | Jet-milled (micronized) proliposomes exhibited excellent aerosol dispersion, drug encapsulation efficiency, and potential as a cost-effective delivery system for targeted TB treatment | Srichana et al. (2022) |
| Tuberculosis | Rifampicin | Spray-dried proliposomes with porous lactose carrier | Proliposomes showed rapid drug release and favourable aerosolization properties (mass median aerodynamic diameter of $6.21 \pm 0.36$ µm and FPF of $9.17 \pm 0.18\%$ with a fast drug release) | Parhizkar et al. (2021) |

(continued)

**Table 4.1** (continued)

| Condition | Medication | Formulation | Main findings | Author and year |
|---|---|---|---|---|
| Tuberculosis | Rifapentine | Spray-dried proliposomes | Drug susceptibility testing revealed that tubercle bacteria is sensitive to the formulation at 10 μg/mL; the formulation was also safe in A549 cells and in vivo at 1–5 mg/kg | Pokharkar et al. (2014) |
| Tuberculosis | Rifapentine | Spray-dried proliposomes (QbD approach) | The optimized proliposomes demonstrated prolonged drug retention in the lung | Patil-Gadhe and Pokharkar (2013) |
| Tuberculosis | Pyrazinamide | Spray-dried proliposomes | Pyrazinamide-proliposomes exhibited favourable aerosolization, reduced toxicity, and higher efficacy against *M. bovis* compared to free pyrazinamide | Rojanarat et al. (2012a) |
| Tuberculosis | Levofloxacin | Spray-dried proliposomes | Levofloxacin-proliposomes demonstrated reduced toxicity, improved aerosolization, and enhanced efficacy against *M. tuberculosis* compared to free levofloxacin | Rojanarat et al. (2012b) |
| Tuberculosis | Isoniazid | Spray-dried proliposomes | Isoniazid-proliposomes showed enhanced pulmonary delivery, reduced systemic toxicity, and higher efficacy against *M. bovis* in alveolar macrophages in vitro | Rojanarat et al. (2011) |
| *P. aeruginosa* | Tobramycin and clarithromycin | Combination proliposome formulation (TOB/CLA-CPROLips) containing tobramycin and clarithromycin | The formulation was stable after 3 months of storage at 60% relative humidity and 25 °C, and showed synergistic antimicrobial activity against *P. aeruginosa* in vitro | Ye et al. (2018) |

(continued)

**Table 4.1** (continued)

| Condition | Medication | Formulation | Main findings | Author and year |
|---|---|---|---|---|
| Fungal infections | Amphotericin B (AmB) | Inhalable proliposomal microparticles/nanoparticles using DPPC and DPPG prepared by spray drying method and delivered using DPI | Smooth, spherical microparticles/nanoparticles were formed with low residual water content at medium to high spray drying rates and the formulation was safe when tested in vitro | Gomez et al. (2020) |
| Fungal infections | Amphotericin B (AmB) | Chitosan-coated liposomes produced from proliposomes using the ethanol-based method followed by delivery through an air-jet nebulizer into a two-stage impinger | The formulation had high drug loading and has potential to be effectively delivered to the peripheral airways via nebulization | Albasarah et al. (2010) |

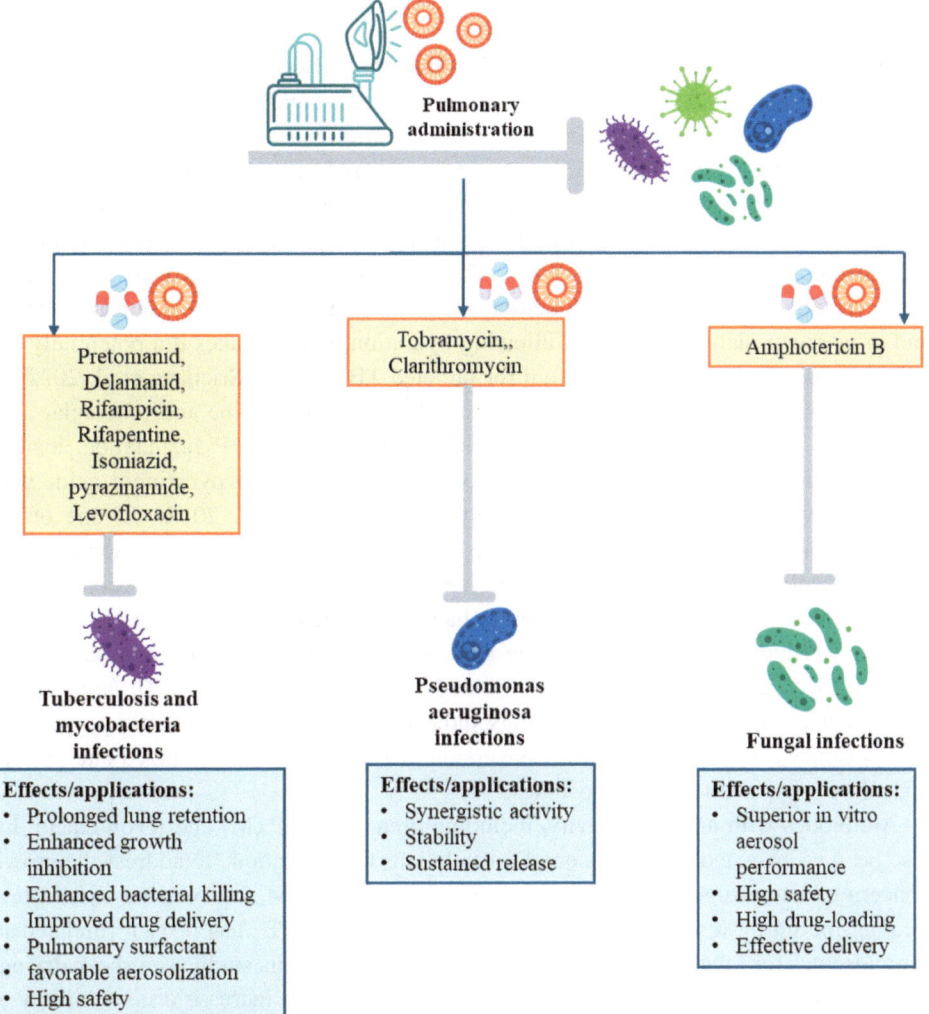

**Fig. 4.2** Therapeutic applications of liposomes and proliposomes delivered via pulmonary administration for treating respiratory tract infections

was incorporated into proliposomes via the same technique (i.e. spray drying) (Patil-Gadhe and Pokharkar 2013). Upon hydration, the formulation was found to have excellent aerosolization properties in vitro upon hydration, exhibiting sustained release of rifapentine, and a longer retention time in the lungs when tested in vivo in an albino rat model (Patil-Gadhe and Pokharkar 2013). This formulation was further explored in another study, which confirmed the safety of these proliposomes both in vitro and in vivo (Pokharkar et al. 2014). Notably, safety was demonstrated at lower doses; therefore, further research is warranted to establish the safety of this formulation upon increasing the drug dose (Pokharkar et al. 2014).

In addition to rifampicin, isoniazid, a first-line anti-TB drug, has also been explored in proliposomal formulations. Rojanarat et al. formulated isoniazid proliposome powders for inhalation, which, upon hydration, were reported to offer enhanced pulmonary delivery, reduced toxicity to respiratory cell lines, and increased efficacy against *Mycobacterium bovis* in infected alveolar macrophages compared with the free drug (Rojanarat et al. 2011). In addition, a novel jet-milling approach for preparing dry proliposome powders for inhalation containing isoniazid and rifampicin was explored (Srichana et al. 2022). This formulation was reported to demonstrate excellent aerosol dispersion performance and drug encapsulation efficiency following hydration, which makes it a potentially cost-effective and efficient delivery system for targeted TB therapy (Srichana et al. 2022).

In addition to isoniazid and rifampicin, another studied first-line anti-tubercular agent, pyrazinamide, in proliposomal formulations has been studied. Pyrazinamide, a nicoti-namide analogue, is a prodrug that is converted in vivo into pyrazinoic acid, which interferes with mycobacterial fatty acid synthase (Zhang et al. 2014). In one of their studies, Rojanarat and co-workers (2012a) prepared pyrazinamide-proliposomes for tar-geting alveolar macrophages in TB. The study demonstrated that, compared with free pyrazinamide, pyrazinamide-proliposomes had favourable aerosolization properties (mass median aerodynamic diameters ranging from 4.26 to 4.39 $\mu$m, with a fine particle frac-tion (FPF) of 20–30%, reduced toxicity, and greater efficacy against *M. bovis* in vitro (Rojanarat et al. 2012a). These findings indicate that the thermodynamic properties of proliposomes can still preserve the conversion of prodrugs at the localized site of action after they are converted into liposomes.

Antibiotics with anti-TB activity, including pretomanid (PTM) and levofloxacin, have also been investigated in proliposomal formulations for inhalation. To address the growing concern of multidrug-resistant tuberculosis (MDR-TB), PTM, a promising candidate for treating multidrug-resistant and extensively drug-resistant TB, has been formulated into proliposomes for inhalation (Aekwattanaphol et al. 2024). This study used spray drying to develop PTM proliposomes with L-leucine and trehalose to increase drug solubility and improve powder aerosolization upon reconstitution in simulated physiological lung fluid (Aekwattanaphol et al. 2024). Compared with PTM alone, the formulations (coded as M1 and M2) were found to have a high drug content, favorable aerosolization properties, and significantly greater antimycobacterial activity against *M. bovis* (Aekwattanaphol et al. 2024). As shown in Fig. 4.3, formulations M1 and M2 had lower minimum inhibitory concentrations (MICs) and minimum bactericidal concentrations (MBCs) compared to PTM alone (Aekwattanaphol et al. 2024). Additionally, liposomal formulations, partic-ularly M2, significantly reduced bacterial viability over 5 days (Aekwattanaphol et al. 2024). On the other hand, levofloxacin, a fluoroquinolone antibiotic, has been incorpo-rated into proliposomes to improve its pulmonary delivery and efficacy in treating TB (Rojanarat et al. 2012b). Compared with free levofloxacin, the levofloxacin-proliposome

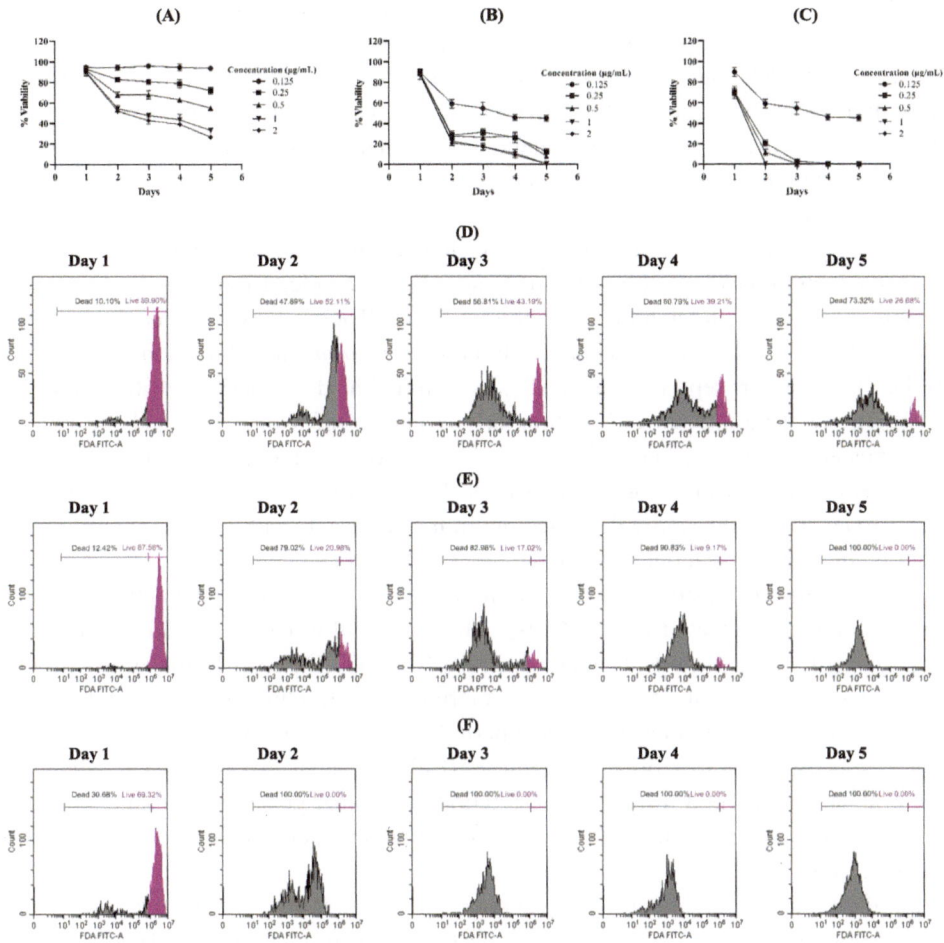

**Fig. 4.3**  The viability of *M. bovis* at various concentrations of **a** PTM, **b** M1 and **c** M2. Flow cytometry histograms showing the fluorescence count versus intensity of viable *M. bovis* treated with 2 μg/mL of PTM in **d** PTM alone, **e** M1 and **f** M2 from day 1 to day 5. Reproduced with permission from Aekwattanaphol et al. (2024). © Elsevier

formulation revealed promising aerosolization properties of liposomes upon hydration, improved safety, and better efficacy against *M. tuberculosis* (Rojanarat et al. 2012b).

### 4.2.1.2   Pseudomonas aeruginosa

*P. aeruginosa* is a gram-negative opportunistic pathogen that has been identified as a major cause of respiratory infections, particularly among patients suffering from cystic fibrosis (Wood et al. 2023). *P. aeruginosa* infections are difficult to treat because of their resistance to antibiotics (Wood et al. 2023). Furthermore, these bacteria can secrete toxins that may damage lung tissue (Wood et al. 2023). To date, numerous studies have validated the therapeutic utility of liposomal and proliposomal formulations of antibiotics via in vitro and in vivo models and in human clinical trials for targeting *P. aeruginosa* infections, particularly in cystic fibrosis patients. The main studies that have reported the successful design and use of proliposomal formulations for potential pulmonary administration and management of *P. aeruginosa* are summarized in Table 4.1, and the main findings are illustrated in Fig. 4.2.

Aminoglycosides have been well investigated for their enhanced antimicrobial activity through pulmonary delivery in liposomes. For example, Bilton et al. investigated the long-term safety, tolerability, and efficacy of amikacin liposome inhalation suspension (ALIS) in cystic fibrosis patients with chronic *P. aeruginosa* infection (Bilton et al. 2021). These authors reported that ALIS was well tolerated and possessed strong antibacterial activity (Bilton et al. 2021). Similarly, Alhariri and Omri evaluated the efficacy of a liposomal bismuth-ethanedithiol-loaded tobramycin formulation (LipoBiEDT-TOB) against *P. aeruginosa* infections (Alhariri and Omri 2013). These findings indicated that LipoBiEDT-TOB successfully disrupted quorum-sensing molecules and virulence factors resulting in reduced bacterial count in the lungs (Alhariri and Omri 2013). Moreover, Halwani et al. conducted a study in which gallium was encapsulated with an aminoglycoside, gentamicin, in liposomes (Lipo-Ga-GEN), which resulted in increased antimicrobial efficacy against *P. aeruginosa* (Halwani et al. 2008). The Lipo-Ga-GEN formulation exhibited notable effects on biofilm eradication and interference with quorum-sensing molecules at minimal concentrations, indicating its superiority over gentamicin monotherapy against antibiotic-resistant strains of *P. aeruginosa* (Halwani et al. 2008). When gentamicin was evaluated alone in a different investigation, liposomal gentamicin formulations displayed superior efficacy compared with free gentamicin in managing pulmonary infections (Mugabe et al. 2005).

Similarly, liposomal suspensions generated upon hydration of aminoglycoside-loaded proliposomes have been investigated for their targeted action against *P. aeruginosa* infection. Ye et al. developed a combination of a proliposome formulation containing both hydrophilic tobramycin and hydrophobic clarithromycin (TOB/CLA-CPROLips) via the core carrier method coupled with spray drying and hydrated them to form liposomes (Ye et al. 2018). This formulation showed potentiated antimicrobial activity against *P. aeruginosa* biofilms and planktonic cells, which could serve as a promising approach for the preparation of proliposomes loaded with a combination of antibiotics (Ye et al. 2018).

In addition to aminoglycosides, liposomal formulations have also been explored as effective carriers of fluoroquinolones (Bandara et al. 2016; Derbali et al. 2019; Haworth et al. 2019; Wang et al. 2024) and macrolides (Alhajlan et al. 2013; Dallal Bashi et al. 2024; Solleti et al. 2015). However, studies validating the benefit of proliposomal formulations of antibiotics from these classes of medication in effectively targeting *P. aeruginosa* are still needed. In the opinion of the authors of this review, because *P. aeruginosa* infection is very difficult to treat, most investigations have focused on finding therapies or excipients that would improve the bactericidal effect of the formulation while addressing issues of formulation stability and scalability (e.g., using proliposomes) was not a priority. This may justify the extensive use of traditional liposomes compared with proliposome-generated liposomes.

### 4.2.1.3 Fungal Infections

The utilization of aerosolized antifungal treatments has appeared to be an encouraging approach for the management and prevention of invasive pulmonary fungal infections, given their capacity for precise administration and diminished systemic adverse effects (Le and Schiller 2010; Vuong et al. 2023). Various studies have investigated the effectiveness, safety, and administration methods of aerosolized liposomal antifungal medications, and the use of proliposome formulations to generate liposomes. Prominent studies that have reported the successful design and use of proliposomal formulations for the pulmonary administration and management of fungal infections are summarized in Table 4.1. Additionally, the main findings of these studies are illustrated in Fig. 4.2.

Amphotericin B is the most studied antifungal agent in aerosolized liposomal formulations because of its effectiveness in the management of invasive fungal infections of the lung. One study focused on the development and characterization of an inhalable proliposomal delivery system containing amphotericin B and lung surfactant-mimic phospholipids, which displayed enhanced aerosol performance and potential for targeted delivery of the drug to the lungs (Gomez et al. 2020). The in vitro aerosol dispersion performance was confirmed to be excellent via the FDA-approved next-generation impactor (NGI) and a human dry powder inhaler (DPI), Handihaler® (Gomez et al. 2020). Smooth, spherical microparticles/nanoparticles were formed with low residual water content at medium to high spray drying rates, and the formulation was safe when tested in vitro (Gomez et al. 2020). In another study, chitosan-coated liposomes prepared via an ethanol-based proliposome method containing amphotericin B (AmB) were developed for delivery via an air-jet nebulizer (Albasarah et al. 2010). The formulation showed significant AmB incorporation, with approximately 80% drug loading (Albasarah et al. 2010). Upon nebulization, AmB was effectively delivered in "respirable" fractions, depositing approximately 60% of AmB in the lower stage of the impinger (Albasarah et al. 2010). Additionally, the antifungal activities against *Candida albicans* and *Candida tropicalis* were similar to those of AmB deoxycholate micelles, with a minimum inhibitory concentration of 0.5 μg/mL (Albasarah et al. 2010). Systemically, liposomes prepared via the proliposome method were also proven superior to traditional liposomal formulations prepared via the thin-film hydration technique. Singodia et al. used a modified ethanol injection method based on proliposomes to produce intravenous amphotericin B liposomes for in situ application (Singodia et al. 2012). The formulation showed high entrapment efficiency (EE) and comparable efficacy to the clinically established amphotericin B liposomes AmBisome (Singodia et al. 2012). These findings suggest a cost-effective approach to the targeted delivery of proliposme-based medications (Singodia et al. 2012). Furthermore, proliposome-based liposomal vesicles incorporating amphotericin B and chemically modified beta-cyclodextrins (beta-CD) significantly improved the survival rate, produced higher antifungal activity, and were 6 times less toxic than free AmB or conventional liposomal AmB (Chakraborty and Naik 2003). Therefore, future research should focus on optimizing and developing proliposomal formulations for combination therapy to improve therapeutic outcomes.

## 4.2.2 Asthma

Asthma represents a significant challenge in terms of therapy owing to its complex biological nature and the lack of effective treatments; asthma manifests as a persistent inflammatory condition that affects the respiratory airways (Papi et al. 2020). Traditional treatments, such as bronchodilators and inhaled corticosteroids, often necessitate frequent adjustments in dosage and are linked to systemic adverse reactions, potentially resulting in patients' non-adherence to the prescribed treatment regimen and inadequate disease management. Liposomal formulations have received much attention recently for the treatment of different airway conditions, including asthma (Mehta et al. 2020; Willis et al. 2012). Remarkably, proliposomes serve as promising formulations to generate vesicles, offering an opportunity to address the limitations of liposomal formulations in drug delivery for the treatment of asthma (Dhiman et al. 2022; Willis et al. 2012). The main studies that have reported the successful design and use of proliposomal formulations for potential pulmonary administration and management of asthma are summarized in Table 4.2.

Salbutamol is a commonly used B2 agonist bronchodilator in the treatment of asthma and is prone to solubility limitations. Therefore, a variety of proliposome compositions encapsulating salbutamol have been designed and processed through various techniques. For example, Omer et al. used the spray drying method to manufacture inhalable proliposome microparticles containing salbutamol sulphate (Omer et al. 2018). The study highlighted the importance of carrier type (lactose monohydrate or mannitol) and lipid-to-carrier ratio, in influencing the FPF and drug EE (Omer et al. 2018). In particular, compared with lactose monohydrate carriers, mannitol-based proliposomes presented better flow properties and FPF values (Omer et al. 2018). In contrast, the drug EE was greater for liposomes generated from lactose monohydrate-based proliposomes (Omer et al. 2018). These findings indicate that spray drying can produce inhalable proliposome microparticles potentially suitable for pulmonary delivery and that the FPF and EE are formulation dependent. Similarly, Al-Najjar and Ghareeb investigated the use of various core carriers (sucrose, mannitol, trehalose, and lactose) in spray-dried proliposomes containing salbutamol sulphate (El-Saadony et al. 2023). They reported that the carrier type significantly affected the properties of the generated liposomes (Al-Najjar and Ghareeb 2020). Furthermore, Elhissi et al. explored the effects of nebulizer design

**Table 4.2** Summary of studies on the pulmonary delivery of liposomal and proliposomal formulations for asthma treatment

| Medication name | Formulation | Device | Main findings | Author and year |
|---|---|---|---|---|
| Beclometasone dipropionate | Spray-dried proliposome powder formulations with different lactose carriers and dispersion media | Dry powder inhaler delivered to a next generation impactor | The combination of water and ethanol as the dispersion medium resulted in superior formulation properties for pulmonary drug delivery, irrespective of the carrier type | Khan et al. (2023) |
| Beclometasone dipropionate | Proliposome powder and tablet formulations Pari-LC Sprint nebulizer was used to generate liposomes from the above formulations | A Pari-LC Sprint air-jet nebulizer and a two-stage impinger as an in vitro aerosol deposition apparatus | Proliposome tablets showed shorter sputtering times and greater drug delivery to the lungs compared to powder formulations | Khan et al. (2021) |
| Salbutamol sulphate | Spray-dried proliposomes with different core carriers (sucrose, mannitol, trehalose, and lactose) | Designed for pulmonary delivery (no specific mention of a final inhalation device) | Carrier type significantly affected the properties of the generated liposomes. Simulated lung fluid was preferable to deionized water for liposome evaluation | Al-Najjar and Ghareeb (2020) |
| Salbutamol sulphate | Spray-dried proliposome microparticles with lactose monohydrate or mannitol as carriers | Dry powder inhaler and a two-stage impinger as an in vitro aerosol deposition apparatus | The FPF and drug EE were influenced by carrier type (mannitol or LMH) and lipid to carrier ratio. Mannitol-based formulations showed higher FPF, while LMH-based formulations offered higher drug EE | Omer et al. (2018) |

(continued)

**Table 4.2** (continued)

| Medication name | Formulation | Device | Main findings | Author and year |
|---|---|---|---|---|
| Beclometasone dipropionate | Proliposome powders with lactose monohydrate, sorbitol, or D-mannitol as carriers | Designed for pulmonary delivery (no specific mention of a final inhalation device) | Carrier type and dispersion medium influenced drug entrapment | Khan et al. (2018) |
| Beclometasone dipropionate | Slurry method for proliposome preparation | Designed for pulmonary delivery (no specific mention of a final inhalation device) | The slurry method for proliposome preparation resulted in smaller liposomes with significantly higher beclometasone EE compared to conventional proliposome | Khan et al. (2015) |
| Beclometasone dipropionate | Hydrogenated soya phosphatidylcholine nanoliposome powders with sucrose as carrier by fluid-bed coating, high-pressure homogenization, and freeze-drying | Designed for pulmonary delivery (no specific mention of a final inhalation device) | Stable nanovesicles with high drug EE (57–69.5%) were produced. Sucrose acted as both a carrier and a cryoprotectant | Gala et al. (2015) |
| Salbutamol sulphate | Ethanol-based proliposomes | Aeroneb Pro vibrating-mesh nebulizer | Liposomes improved nebulized drug output and FPF compared to conventional drug solutions (FPF values 57.85% and 45.81%, respectively). The improved performance was attributed to the reduction of surface tension by phospholipids | Elhissi et al. (2013) |

(continued)

**Table 4.2** (continued)

| Medication name | Formulation | Device | Main findings | Author and year |
|---|---|---|---|---|
| NA (phospholipids only) | Particulate-based proliposomes with sucrose carrier particles coated with phospholipids | Pari LC Plus nebulizer and a two-stage impinger as an in vitro aerosol deposition apparatus | All proliposome formulations produced high aerosol and phospholipid outputs and were delivered in high fractions to the lower stage of a two-stage impinger | Elhissi et al. (2012) |
| Salbutamol sulphate | Ethanol-based proliposomes hydrated with NaCl or sucrose solutions | Aeroneb Go vibrating-mesh nebulizer | The choice of hydration medium and nebulizer design influenced aerosol performance. NaCl solution as the dispersion medium resulted in generation of smaller droplet size and higher aerosol mass and phospholipid outputs | Elhissi et al. (2011) |
| Salbutamol sulphate | Multilamellar and oligolamellar liposomes from ethanol-based proliposomes | Pari LC Plus air-jet nebulizer Aeroneb Pro vibrating-mesh nebulizer Omron NE U22 vibrating-mesh nebulizer | Liposomes generated from proliposomes can be aerosolized in small droplets using various nebulizers. Vibrating-mesh nebulizers showed greater dependence on formulation parameters compared to jet nebulizers | Elhissi et al. (2006) |

and hydration medium viscosity on the aerosolization of liposomes containing salbutamol sulphate (Elhissi et al. 2011). The study demonstrated that lower viscosity hydration media and optimized nebulizer designs led to improved aerosol mass and phospholipid outputs (Elhissi et al. 2011). The performance of air-jet nebulizers and vibrating-mesh nebulizers was also compared in aerosolizing liposomes generated from ethanol-based proliposomes (Elhissi et al. 2006). Both nebulizer types produced small aerosol droplets, but vibrating-mesh nebulizers showed greater dependence on formulation parameters (Elhissi et al. 2006). Other investigations on liposomes prepared via ethanol-based proliposomes revealed that ethanol-based proliposomes containing salbutamol sulphate, when hydrated and aerosolized by vibrating-mesh nebulizers, resulted in improved aerosol output and FPF compared with those of the corresponding drug solutions (Elhissi et al. 2013). Additionally, the choice of hydration medium and nebulizer design significantly affects aerosol performance (Elhissi et al. 2013). All these findings are expected to affect the in vivo outcome, which should constitute a major part the future investigations.

On the other hand, proliposomes have been utilized for the encapsulation of inhaled corticosteroids (ICSs) as a promising approach to increase therapeutic effectiveness and reduce the possibility of adverse effects. For example, Khan et al. developed and compared proliposome powder and tablet formulations for delivering beclometasone dipropionate liposomes via a Pari-LC Sprint nebulizer (Khan et al. 2021). The study revealed that proliposome tablets, upon disintegration and dispersion in water, offered advantages in terms of increasing the output of the nebulizer, thus offering greater potential for pulmonary delivery (Khan et al. 2021). In addition, Gala et al. utilized fluid-bed coating, high-pressure homogenization, and freeze-drying to produce beclometasone dipropionate-loaded nanoliposome powders, as demonstrated in Fig. 4.4 (Gala et al. 2015). This approach demonstrated a scalable method for producing stable nanovesicles with high drug EE (Gala et al. 2015). In another exploration, beclometasone was prepared via a novel "slurry method" for preparing proliposome powders, resulting in liposomes with a smaller median size and greater beclometasone EE than traditional proliposome and thin-film hydration methods do (Khan et al. 2015). The influence of carbohydrate carriers and separation conditions was explored and found to play a role in the entrapment of beclometasone dipropionate in liposomes generated from proliposomes (Khan et al. 2018). In particular, for deuterium oxide ($D_2O$), the entrapment efficiency was $19.69 \pm 5.88\%$ for lactose monohydrate (LMH)-based proliposomes, $28.78 \pm 4.69\%$ for sorbitol-based proliposomes, and $34.84 \pm 3.62\%$ for D-mannitol-based proliposomes, and the use of $D_2O$ was found to be essential for accurate determination of steroid entrapment in the vesicles (Khan et al. 2018). Additionally, Khan et al. reported that the use of a mixture of water and ethanol as the dispersion medium resulted in superior formulation properties for dry powder aerosol performance, which highlights the impact of the dispersion medium and formulation optimization (Khan et al. 2023).

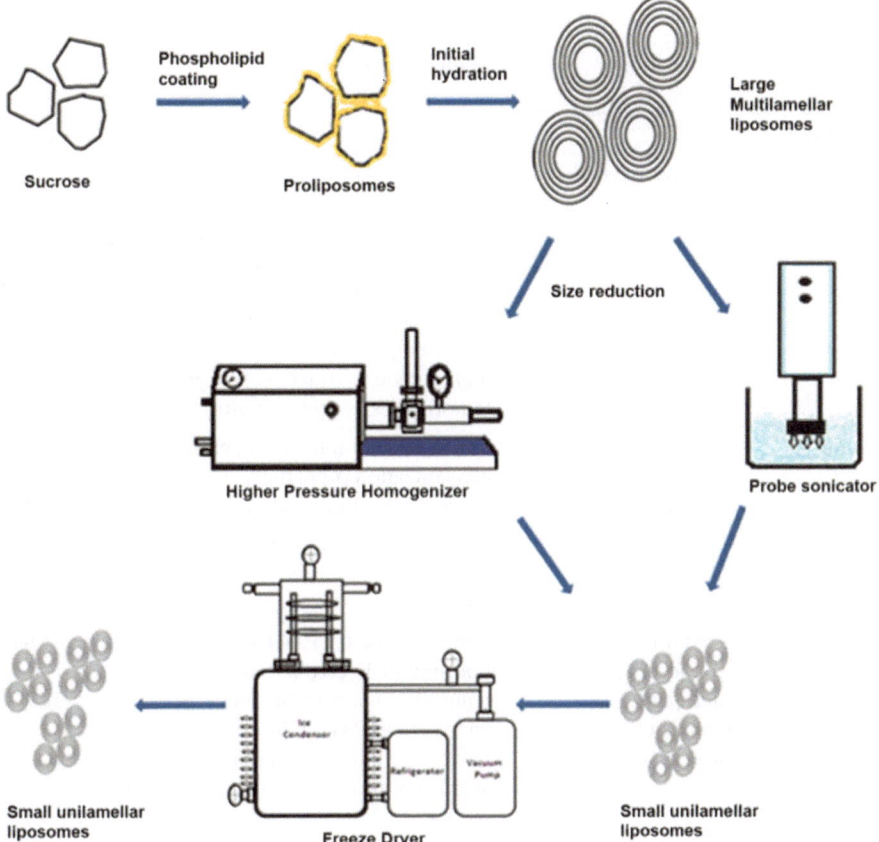

**Fig. 4.4** The large-scale production method employed by Gala et al. (2015). Fluid-bed coating, high-pressure homogenization, and freeze-drying were performed to produce dipropionate-loaded nanoliposome powders. Reproduced with permission from Gala et al. (2015). © Elsevier

### 4.2.3   Lung Cancer

Lung cancer is a significant contributor to cancer-related mortality worldwide, primarily because of its aggressive nature and the barriers related to treatment effectiveness (Dela Cruz et al. 2011). Although conventional chemotherapy and radiotherapy are well established for the treatment of lung cancer, they have several limitations, such as systemic toxicity, low drug bioavailability, and the emergence of multidrug resistance (Eslami et al. 2024; Huang et al. 2017). The emergence of nanotechnology, including liposomes, serves as an alternative modality to assist in overcoming the limitations of current lung cancer therapies (Chehelgerdi et al. 2023). The main studies that have reported the successful

**Table 4.3** Summary of studies on the pulmonary delivery of liposomal and proliposomal formulations for lung cancer treatment and gene therapy

| Medication name | Formulation | Main findings | Author and year |
|---|---|---|---|
| *Lung cancer* | | | |
| Paclitaxel | Proliposome tablets | One of the formulations made from soya phosphatidylcholine and lactose monohydrate carrier with lipid to carrier ratio of 1:15 w/w complied with British Pharmacopeia quality standards. The ultrasonic nebulization showed better performance. PTX-loaded formulation was highly toxic to cancer cells (type) but safe to normal cells (type) in vitro | Khan et al. (2020) |
| Curcumin | Nano-spray drying to prepare proliposomes using hydroxypropyl beta-cyclodextrin as a carrier | The formulation led to superior aerosolization properties, enhanced lung tissue penetration, and improved pharmacokinetic profile | Adel et al. (2021) |
| *Gene therapy* | | | |
| CYP1A1 gene silencing | Lipid film-coated proliposome microparticles administered intratumorally | Inhibiting the CYP1A1 gene using cationic liposome-delivered siRNA effectively reduced tumor growth and triggered apoptosis in PAH-induced lung cancer models both in vitro and in vivo. Further details are provided in the subsequent section | Zhang et al. (2019) |

design and use of proliposomal formulations for potential pulmonary administration and treatment of lung cancer are summarized in Table 4.3.

The formulation of proliposome preparations for inhalable therapies represents a promising strategy for the treatment of lung cancer, as it enables the direct delivery of high concentrations of medications to the lungs (Willis et al. 2012). Khan et al. designed paclitaxel-loaded proliposome tablets for nebulization against lung cancer (Khan et al. 2020). In this study, multiple proliposome powders (labelled F1–27) were prepared to incorporate paclitaxel, which was subsequently formulated into dry powders incorporating paclitaxel (Khan et al. 2020). Depending on the formulation, proliposomal tablets demonstrated consistent weight, good mechanical strength, appropriate thickness, and rapid disintegration, meeting the British Pharmacopeia standards, with soya phosphatidylcholine and lactose monohydrate carrier at a lipid-to-carrier ratio of 1:15 w/w being most successful in meeting the standards (Khan et al. 2020). The tablet formulation was hydrated in a nebulizer reservoir to form a liposome suspension. When nebulized, it performed better with an ultrasonic nebulizer, resulting in a shorter nebulization time and higher output than those of a vibrating mesh nebulizer (Khan et al. 2020). The paclitaxel-loaded tablets were hydrated with deionized water at a concentration of 10% for the in vitro cytotoxicity test and were found to be more toxic to the MRC-5 SV2 lung cancer cells in vitro than to the normal MRC-5 cell line (Khan et al. 2020). These findings provide a promising foundation for further research into the use of proliposomal formulations for safe and effective pulmonary delivery of anticancer drugs in the treatment of lung cancer.

In addition to traditional anticancer medications, proliposomes have also been explored for the delivery of other investigational therapeutic agents to the lungs. For example, Adel et al. developed proliposomes encapsulating curcumin, a polyphenolic compound with anti-inflammatory and anticancer properties (Adel et al. 2021). Compared with curcumin powder alone, these proliposomes, which were prepared via nano spray drying, showed superior aerosolization and lung deposition in vivo using an albino rat animal model, with improved inhibition of lung tumour cells using A549 cell lines (Adel et al. 2021). As depicted in Fig. 4.5, the lung concentration–time profile of curcumin in rats was evaluated after intratracheal administration of free curcumin powder and proliposomal formulations (Adel et al. 2021). Compared with the free drug, the proliposomal formulations presented improved pharmacokinetic parameters, including a shorter time to peak concentration ($T_{max}$), higher peak concentration ($C_{max}$), and increased area under the curve (AUC). These findings suggest that proliposomal formulations can increase the rate and extent of curcumin uptake by lung tissue (Adel et al. 2021).

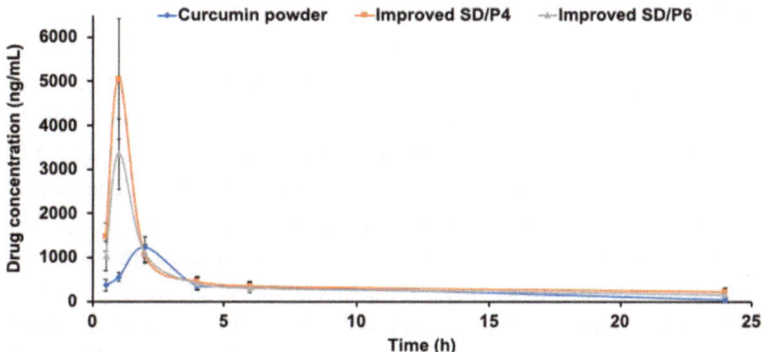

**Fig. 4.5** Lung concentration–time curves for the curcumin-free powder and curcumin proliposomal formulations. Reproduced from Adel et al. (2021), originally published by and used with permission from Dove Medical Press Ltd.

### 4.2.4  Gene Therapy

Gene therapy has been identified as an innovative strategy for the management of a broad range of genetic disorders, such as those that affect pulmonary function (Bañuls et al. 2020; Chow et al. 2020). Nonetheless, effective transportation of therapeutic genes to specific cells poses a notable barrier, particularly with respect to achieving efficient transfection and reducing adverse effects (Baliga and Dean 2021; Kim et al. 2016). Liposomes, which serve as nonviral vectors for gene delivery, present practical solutions owing to their compatibility with biological systems and ability to deliver genes to targeted sites (Patil et al. 2019; Wang et al. 2023). Liposomal formulations can be customized to augment gene expression, alleviate immune reactions, and sustain therapeutic outcomes.

Numerous studies have investigated and established the therapeutic potential of liposomes in pulmonary gene therapy for various medical conditions. These investigations included cystic fibrosis (Davies et al. 2014; McLachlan et al. 2011), idiopathic pulmonary fibrosis (Wang et al. 2021; Yan et al. 2023), lung cancer (Allon et al. 2012; Dames et al. 2007; Li et al. 2011; Zou et al. 2000), and pulmonary hypertension (Gong et al. 2005). Zhang et al. used an RNA interference (RNAi) approach to suppress CYP1A1 overexpression by delivering CYP1A1 siRNA via cationic liposomes (Zhang et al. 2019) (Table 4.3). These liposomes were prepared by coating sorbitol particles with 1,2-dioleoyl-3-trimethylammonium-propane (DOTAP) and dioleoyl-167 phosphatidylethanolamine (DOPE) in a chloroform solution via a modified rotary evaporator with a feedline tube, followed by solvent evaporation to form proliposome granules (Zhang et al. 2019). They were then hydrated, sonicated, and mixed with siRNA to create cationic liposome–siRNA complexes (Zhang et al. 2019). Treatment of polycyclic aromatic hydrocarbon (PAH)-induced human alveolar adenocarcinoma cells with these siRNA-loaded liposomes led to a reduction in CYP1A1 mRNA and protein levels and enzymatic activity, which induced apoptosis and prevented the formation of multicellular tumour spheroids in vitro (Zhang et al. 2019). To evaluate the effect of CYP1A1 silencing on tumour progression in vivo, the antitumor activity of CL-siRNA in an A549 xenograft nude mouse model was assessed (Fig. 4.6a) (Zhang et al. 2019). The results indicated that the tumour growth rate in mice treated with cationic liposomes or Lipo2000 was significantly slower than that in the control groups, which included animals receiving PBS or naked (free) siRNA injections (Fig. 4.6b, c) (Zhang et al. 2019). Additionally, RT–PCR analysis on the third day post-intratumoral administration revealed downregulation of CYP1A1 gene expression in the tumours (Fig. 4.6d) (Zhang et al. 2019). Overall, this study demonstrated that liposome-based gene delivery is a promising method for targeting oncogenes such as CYP1A1, potentially offering a new therapeutic strategy for lung cancer treatment. Importantly, this study was conducted through injecting the tumor with the formulation, and future investigations should involve the delivery of the formulation either in the form of aerosol or through intratracheal instillation.

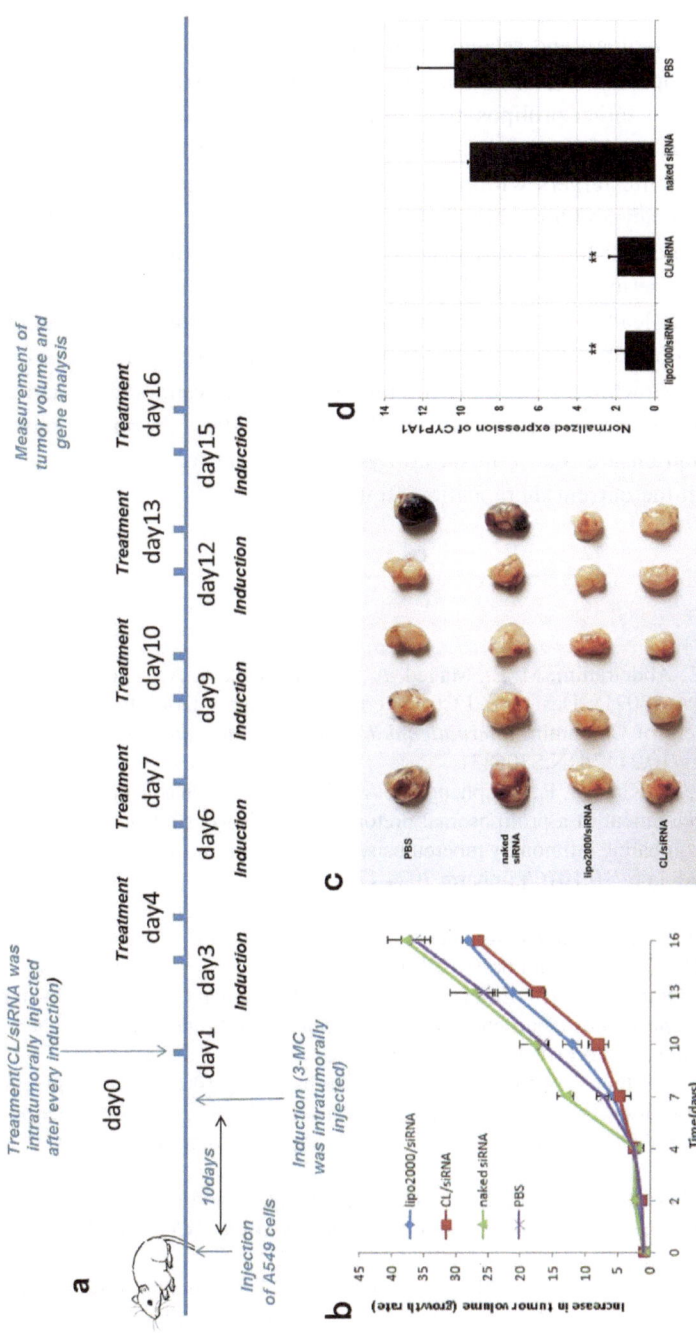

**Fig. 4.6** Inhibition of tumour growth via the use of CYP1A1-siRNA in a mouse model system. BALB/c-nude mice were injected with 10 million A549 cells in serum-free medium subcutaneously into the right flank. The tumour bearing mice were divided into four treatment groups (n = 5). **a** All the mice were injected with 3-MC, and after 10 days, they were treated with (i) PBS, (ii) CL-CYP1A1-siRNA, (iii) naked-siRNA or (iv) lipofectamine-complexed CYP1A1-siRNA. **b** The volume of each tumour was measured at the indicated time points as described in the methods section. The results are expressed as the means (n = 5) ± SDs. **c** Mice were sacrificed after 18 days with six intratumoral injection of CYP1A1 siRNA, and images of each tumour were taken as shown (n = 5). **d** Total RNA was isolated from the tumours of each mouse. The expression of the CYP1A1 gene was quantified via RT-PCR (data are expressed as the means ± SD; n = 3). Reproduced with permission from Zhang et al. (2019). © Elsevier

## 4.3    Conclusion

With the growth of nanotechnology and related fields, various proliposome formulations have been designed, developed, and investigated as drug delivery carriers, particularly for pulmonary applications, as either proliposomes, or following hydration, as liposomes. Through encapsulation, various bioactive molecules and conventional drugs have been effectively delivered to specific targets within pulmonary airways to treat multiple illnesses. In particular, the efficacy and tolerability of proliposomal-based formulations have been remarkably demonstrated for tuberculosis, *P. aeruginosa* infection, fungal infections, asthma, lung cancer, and gene therapy. Overall, the use of proliposomal formulations in pulmonary drug delivery can lead to significant developments in the efficient treatment of a variety of diseases by improving therapeutic outcomes, safety profiles, patient convenience, and cost-effectiveness while overcoming the limitations of conventional therapies and liposomes. Additional research is warranted to further improve these promising pharmaceutical preparations and ensure their clinical utility in addressing the challenges and limitations associated with the current therapeutic options.

## References

Adel, I. M., ElMeligy, M. F., Abdelrahim, M. E., Maged, A., Abdelkhalek, A. A., Abdelmoteleb, A. M., & Elkasabgy, N. A. (2021). Design and Characterization of Spray-Dried Proliposomes for the Pulmonary Delivery of Curcumin. *International Journal of Nanomedicine, Volume 16*, 2667–2687. https://doi.org/10.2147/IJN.S306831

Aekwattanaphol, N., Das, S. C., Khadka, P., Nakpheng, T., Ali Khumaini Mudhar Bintang, M., & Srichana, T. (2024). Development of a proliposomal pretomanid dry powder inhaler as a novel alternative approach for combating pulmonary tuberculosis. *International Journal of Pharmaceutics, 664*, 124608. https://doi.org/10.1016/j.ijpharm.2024.124608

Albasarah, Yacoub, Somavarapu, Satyanarayana, Stapleton, Paul, & Taylor, Kevin. (2010). Chitosan-coated antifungal formulations for nebulisation. *Journal of Pharmacy and Pharmacology, 62*(7), 821–828. https://doi.org/10.1211/jpp.62.05.0002

Alhajlan, M., Alhariri, M., & Omri, A. (2013). Efficacy and Safety of Liposomal Clarithromycin and Its Effect on *Pseudomonas aeruginosa* Virulence Factors. *Antimicrobial Agents and Chemotherapy, 57*(6), 2694–2704. https://doi.org/10.1128/AAC.00235-13

Alhariri, M., & Omri, A. (2013). Efficacy of Liposomal Bismuth-Ethanedithiol-Loaded Tobramycin after Intratracheal Administration in Rats with Pulmonary *Pseudomonas aeruginosa* Infection. *Antimicrobial Agents and Chemotherapy, 57*(1), 569–578. https://doi.org/10.1128/AAC.016 34-12

Allon, N., Saxena, A., Chambers, C., & Doctor, B. P. (2012). A new liposome-based gene delivery system targeting lung epithelial cells using endothelin antagonist. *Journal of Controlled Release, 160*(2), 217–224. https://doi.org/10.1016/j.jconrel.2011.10.033

Al-Najjar, Basma, & Ghareeb, Mowafaq. (2020). Preparation of Inhalable Salbutamol Proliposome Using Different Core Carriers and Investigate the Impact of Simulated Lung Fluid Use in Evaluation. *Systematic Reviews in Pharmacy, 11*(4), 629–639.

Alsayed, S. S. R., & Gunosewoyo, H. (2023). Tuberculosis: Pathogenesis, Current Treatment Regimens and New Drug Targets. *International Journal of Molecular Sciences*, *24*(6), 5202. https://doi.org/10.3390/ijms24065202

Baliga, U. K., & Dean, D. A. (2021). Pulmonary gene delivery—Realities and possibilities. *Experimental Biology and Medicine*, *246*(3), 260–274. https://doi.org/10.1177/1535370220965985

Bandara, H. M. H. N., Herpin, M. J., Kolacny, D., Harb, A., Romanovicz, D., & Smyth, H. D. C. (2016). Incorporation of Farnesol Significantly Increases the Efficacy of Liposomal Ciprofloxacin against *Pseudomonas aeruginosa* Biofilms *in Vitro*. *Molecular Pharmaceutics*, *13*(8), 2760–2770. https://doi.org/10.1021/acs.molpharmaceut.6b00360

Bañuls, L., Pellicer, D., Castillo, S., Navarro-García, M. M., Magallón, M., González, C., & Dasí, F. (2020). Gene Therapy in Rare Respiratory Diseases: What Have We Learned So Far? *Journal of Clinical Medicine*, *9*(8), 2577. https://doi.org/10.3390/jcm9082577

Bilton, D., Fajac, I., Pressler, T., Clancy, J. P., Sands, D., Minic, P., Cipolli, M., Galeva, I., Solé, A., Quittner, A. L., Jumadilova, Z., Ciesielska, M., & Konstan, M. W. (2021). Long-term amikacin liposome inhalation suspension in cystic fibrosis patients with chronic *P. aeruginosa* infection. *Journal of Cystic Fibrosis*, *20*(6), 1010–1017. https://doi.org/10.1016/j.jcf.2021.05.013

Chakraborty K, & Naik S. (2003). Therapeutic and hemolytic evaluation of in-situ liposomal preparation containing amphotericin - beta complexed with different chemically modified beta - cyclodextrins. *J Pharm Sci*, *6*(3), 231–237.

Chehelgerdi, M., Chehelgerdi, M., Allela, O. Q. B., Pecho, R. D. C., Jayasankar, N., Rao, D. P., Thamaraikani, T., Vasanthan, M., Viktor, P., Lakshmaiya, N., Saadh, M. J., Amajd, A., Abo-Zaid, M. A., Castillo-Acobo, R. Y., Ismail, A. H., Amin, A. H., & Akhavan-Sigari, R. (2023). Progressing nanotechnology to improve targeted cancer treatment: overcoming hurdles in its clinical implementation. *Molecular Cancer*, *22*(1), 169. https://doi.org/10.1186/s12943-023-01865-0

Choudhary, M., Chaurawal, N., Barkat, Md. A., & Raza, K. (2022). Proliposome-Based Nanostrategies: Challenges and Development as Drug Delivery Systems. *AAPS PharmSciTech*, *23*(8), 293. https://doi.org/10.1208/s12249-022-02443-1

Chow, M. Y. T., Qiu, Y., & Lam, J. K. W. (2020). Inhaled RNA Therapy: From Promise to Reality. *Trends in Pharmacological Sciences*, *41*(10), 715–729. https://doi.org/10.1016/j.tips.2020.08.002

Dallal Bashi, Y. H., Ali, A., Al Ayoub, Y., Assi, K. H., Mairs, R., McCarthy, H. O., Tunney, M. M., & Kett, V. L. (2024). Inhaled dry powder liposomal azithromycin for treatment of chronic lower respiratory tract infection. *International Journal of Pharmaceutics*, *653*, 123841. https://doi.org/10.1016/j.ijpharm.2024.123841

Dames, P., Laner, A., Maucksch, C., Aneja, M. K., & Rudolph, C. (2007). Targeting of the glucocorticoid hormone receptor with plasmid DNA comprising glucocorticoid response elements improves nonviral gene transfer efficiency in the lungs of mice. *The Journal of Gene Medicine*, *9*(9), 820–829. https://doi.org/10.1002/jgm.1082

Davies, L. A., Nunez-Alonso, G. A., McLachlan, G., Hyde, S. C., & Gill, D. R. (2014). Aerosol Delivery of DNA/Liposomes to the Lung for Cystic Fibrosis Gene Therapy. *Human Gene Therapy Clinical Development*, *25*(2), 97–107. https://doi.org/10.1089/humc.2014.019

Dela Cruz, C. S., Tanoue, L. T., & Matthay, R. A. (2011). Lung Cancer: Epidemiology, Etiology, and Prevention. *Clinics in Chest Medicine*, *32*(4), 605–644. https://doi.org/10.1016/j.ccm.2011.09.001

Derbali, R. M., Aoun, V., Moussa, G., Frei, G., Tehrani, S. F., Del'Orto, J. C., Hildgen, P., Roullin, V. G., & Chain, J. L. (2019). Tailored Nanocarriers for the Pulmonary Delivery of Levofloxacin against *Pseudomonas aeruginosa*: A Comparative Study. *Molecular Pharmaceutics*, *16*(5), 1906–1916. https://doi.org/10.1021/acs.molpharmaceut.8b01256

Dhiman, N., Sarvaiya, J., & Mohindroo, P. (2022). A drift on liposomes to proliposomes: recent advances and promising approaches. *Journal of Liposome Research*, *32*(4), 317–331. https://doi. org/10.1080/08982104.2021.2019762

Elhissi, A. (2017). Liposomes for Pulmonary Drug Delivery: The Role of Formulation and Inhalation Device Design. *Current Pharmaceutical Design*, *23*(3), 362–372. https://doi.org/10.2174/138161 2823666161116114732

Elhissi, A., Karnam, K., Danesh-Azari, M.-R., Gill, H., & Taylor, K. (2006). Formulations generated from ethanol-based proliposomes for delivery via medical nebulizers. *Journal of Pharmacy and Pharmacology*, *58*(7), 887–894. https://doi.org/10.1211/jpp.58.7.0002

Elhissi, A., Gill, H., Ahmed, W., & Taylor, K. (2011). Vibrating-mesh nebulization of liposomes generated using an ethanol-based proliposome technology. *Journal of Liposome Research*, *21*(2), 173–180. https://doi.org/10.3109/08982104.2010.505574

Elhissi, A., Ahmed, W., & Taylor, K. M. G. (2012). Laser Diffraction and Electron Microscopy Studies on Inhalable Liposomes Generated from Particulate-Based Proliposomes Within a Medical Nebulizer. *Journal of Nanoscience and Nanotechnology*, *12*(8), 6693–6699. https://doi.org/10. 1166/jnn.2012.4566

Elhissi, Brar, Jasmeet, Najlah, Mohammad, Roberts, Simon, Faheem, Ahmed, & Taylor, Kevin. (2013). An Ethanol-Based Proliposome Technology for Enhanced Delivery and Improved "Respirability" of Antiasthma Aerosols Generated Using a Micropump Vibrating-Mesh Nebulizer. *Journal of Pharmaceutical Technology, Research and Management*, *1*(2), 171–180. https://doi. org/10.15415/jptrm.2013.12010

El-Saadony, M. T., Yang, T., Korma, S. A., Sitohy, M., Abd El-Mageed, T. A., Selim, S., Al Jaouni, S. K., Salem, H. M., Mahmmod, Y., Soliman, S. M., Mo'men, S. A. A., Mosa, W. F. A., El-Wafai, N. A., Abou-Aly, H. E., Sitohy, B., Abd El-Hack, M. E., El-Tarabily, K. A., & Saad, A. M. (2023). Impacts of turmeric and its principal bioactive curcumin on human health: Pharmaceutical, medicinal, and food applications: A comprehensive review. *Frontiers in Nutrition*, *9*. https:// doi.org/10.3389/fnut.2022.1040259

Eslami, M., Memarsadeghi, O., Davarpanah, A., Arti, A., Nayernia, K., & Behnam, B. (2024). Overcoming Chemotherapy Resistance in Metastatic Cancer: A Comprehensive Review. *Biomedicines*, *12*(1), 183. https://doi.org/10.3390/biomedicines12010183

Ferreira, M., Ogren, M., Dias, J. N. R., Silva, M., Gil, S., Tavares, L., Aires-da-Silva, F., Gaspar, M. M., & Aguiar, S. I. (2021). Liposomes as Antibiotic Delivery Systems: A Promising Nanotechnological Strategy against Antimicrobial Resistance. *Molecules*, *26*(7), 2047. https://doi.org/10. 3390/molecules26072047

Gala, R. P., Khan, I., Elhissi, A. M. A., & Alhnan, M. A. (2015). A comprehensive production method of self-cryoprotected nano-liposome powders. *International Journal of Pharmaceutics*, *486*(1–2), 153–158. https://doi.org/10.1016/j.ijpharm.2015.03.038

Gomez, A. I., Acosta, M. F., Muralidharan, P., Yuan, J. X.-J., Black, S. M., Hayes, D., & Mansour, H. M. (2020). Advanced spray dried proliposomes of amphotericin B lung surfactant-mimic phospholipid microparticles/nanoparticles as dry powder inhalers for targeted pulmonary drug delivery. *Pulmonary Pharmacology & Therapeutics*, *64*, 101975. https://doi.org/10.1016/j.pupt. 2020.101975

Gong, F., Tang, H., Lin, Y., Gu, W., Wang, W., & Kang, M. (2005). Gene transfer of vascular endothelial growth factor reduces bleomycin-induced pulmonary hypertension in immature rabbits. *Pediatrics International*, *47*(3), 242–247. https://doi.org/10.1111/j.1442-200x.2005.02060.x

Halwani, M., Yebio, B., Suntres, Z. E., Alipour, M., Azghani, A. O., & Omri, A. (2008). Co-encapsulation of gallium with gentamicin in liposomes enhances antimicrobial activity of gentamicin against *Pseudomonas aeruginosa*. *Journal of Antimicrobial Chemotherapy*, *62*(6), 1291–1297. https://doi.org/10.1093/jac/dkn422

Haworth, C. S., Bilton, D., Chalmers, J. D., Davis, A. M., Froehlich, J., Gonda, I., Thompson, B., Wanner, A., & O'Donnell, A. E. (2019). Inhaled liposomal ciprofloxacin in patients with non-cystic fibrosis bronchiectasis and chronic lung infection with *Pseudomonas aeruginosa* (ORBIT-3 and ORBIT-4): two phase 3, randomised controlled trials. *The Lancet Respiratory Medicine, 7*(3), 213–226. https://doi.org/10.1016/S2213-2600(18)30427-2

Huang, C.-Y., Ju, D.-T., Chang, C.-F., Muralidhar Reddy, P., & Velmurugan, B. K. (2017). A review on the effects of current chemotherapy drugs and natural agents in treating non–small cell lung cancer. *BioMedicine, 7*(4), 23. https://doi.org/10.1051/bmdcn/2017070423

Huang, Y., Chang, Z., Gao, Y., Ren, C., Lin, Y., Zhang, X., Wu, C., Pan, X., & Huang, Z. (2024). Overcoming the Low-Stability Bottleneck in the Clinical Translation of Liposomal Pressurized Metered-Dose Inhalers: A Shell Stabilization Strategy Inspired by Biomineralization. *International Journal of Molecular Sciences, 25*(6), 3261. https://doi.org/10.3390/ijms25063261

Khan, I., Yousaf, S., Subramanian, S., Korale, O., Alhnan, M. A., Ahmed, W., Taylor, K. M. G., & Elhissi, A. (2015). Proliposome powders prepared using a slurry method for the generation of beclometasone dipropionate liposomes. *International Journal of Pharmaceutics, 496*(2), 342–350. https://doi.org/10.1016/j.ijpharm.2015.10.002

Khan, I., Yousaf, S., Subramanian, S., Alhnan, M. A., Ahmed, W., & Elhissi, A. (2018). Proliposome Powders for the Generation of Liposomes: the Influence of Carbohydrate Carrier and Separation Conditions on Crystallinity and Entrapment of a Model Antiasthma Steroid. *AAPS PharmSciTech, 19*(1), 262–274. https://doi.org/10.1208/s12249-017-0793-2

Khan, I., Lau, K., Bnyan, R., Houacine, C., Roberts, M., Isreb, A., Elhissi, A., & Yousaf, S. (2020). A Facile and Novel Approach to Manufacture Paclitaxel-Loaded Proliposome Tablet Formulations of Micro or Nano Vesicles for Nebulization. *Pharmaceutical Research, 37*(6), 116. https://doi.org/10.1007/s11095-020-02840-w

Khan, I., Yousaf, S., Najlah, M., Ahmed, W., & Elhissi, A. (2021). Proliposome powder or tablets for generating inhalable liposomes using a medical nebulizer. *Journal of Pharmaceutical Investigation, 51*(1), 61–73. https://doi.org/10.1007/s40005-020-00495-8

Khan, I., Al-Hasani, A., Khan, M. H., Khan, A. N., -Alam, F., Sadozai, S. K., Elhissi, A., Khan, J., & Yousaf, S. (2023). Impact of dispersion media and carrier type on spray-dried proliposome powder formulations loaded with beclomethasone dipropionate for their pulmonary drug delivery via a next generation impactor. *PLOS ONE, 18*(3), e0281860. https://doi.org/10.1371/journal.pone.0281860

Kim, N., Duncan, G. A., Hanes, J., & Suk, J. S. (2016). Barriers to inhaled gene therapy of obstructive lung diseases: A review. *Journal of Controlled Release, 240*, 465–488. https://doi.org/10.1016/j.jconrel.2016.05.031

Le, J., & Schiller, D. S. (2010). Aerosolized Delivery of Antifungal Agents. *Current Fungal Infection Reports, 4*(2), 96–102. https://doi.org/10.1007/s12281-010-0011-0

Li, P., Liu, D., Sun, X., Liu, C., Liu, Y., & Zhang, N. (2011). A novel cationic liposome formulation for efficient gene delivery via a pulmonary route. *Nanotechnology, 22*(24), 245104. https://doi.org/10.1088/0957-4484/22/24/245104

McLachlan, G., Davidson, H., Holder, E., Davies, L. A., Pringle, I. A., Sumner-Jones, S. G., Baker, A., Tennant, P., Gordon, C., Vrettou, C., Blundell, R., Hyndman, L., Stevenson, B., Wilson, A., Doherty, A., Shaw, D. J., Coles, R. L., Painter, H., Cheng, S. H., … Collie, D. D. S. (2011). Preclinical evaluation of three non-viral gene transfer agents for cystic fibrosis after aerosol delivery to the ovine lung. *Gene Therapy, 18*(10), 996–1005. https://doi.org/10.1038/gt.2011.55

Mehta, P. P., Ghoshal, D., Pawar, A. P., Kadam, S. S., & Dhapte-Pawar, V. S. (2020). Recent advances in inhalable liposomes for treatment of pulmonary diseases: Concept to clinical stance. *Journal of Drug Delivery Science and Technology, 56*, 101509. https://doi.org/10.1016/j.jddst.2020.101509

Mugabe, C., Azghani, A. O., & Omri, A. (2005). Liposome-mediated gentamicin delivery: development and activity against resistant strains of *Pseudomonas aeruginosa* isolated from cystic fibrosis patients. *Journal of Antimicrobial Chemotherapy, 55*(2), 269–271. https://doi.org/10.1093/jac/dkh518

Omer, H. K., Hussein, N. R., Ferraz, A., Najlah, M., Ahmed, W., Taylor, K. M. G., & Elhissi, A. M. A. (2018). Spray-Dried Proliposome Microparticles for High-Performance Aerosol Delivery Using a Monodose Powder Inhaler. *AAPS PharmSciTech, 19*(5), 2434–2448. https://doi.org/10.1208/s12249-018-1058-4

Papi, A., Blasi, F., Canonica, G. W., Morandi, L., Richeldi, L., & Rossi, A. (2020). Treatment strategies for asthma: reshaping the concept of asthma management. *Allergy, Asthma & Clinical Immunology, 16*(1), 75. https://doi.org/10.1186/s13223-020-00472-8

Parhizkar, E., Sadeghinia, D., Hamishehkar, H., Yaqoubi, S., Nokhodchi, A., & Alipour, S. (2021). Carrier Effect in Development of Rifampin Loaded Proliposome for Pulmonary Delivery: A Quality by Design Study. *Advanced Pharmaceutical Bulletin.* https://doi.org/10.34172/apb.2022.032

Patil, S., Gao, Y.-G., Lin, X., Li, Y., Dang, K., Tian, Y., Zhang, W.-J., Jiang, S.-F., Qadir, A., & Qian, A.-R. (2019). The Development of Functional Non-Viral Vectors for Gene Delivery. *International Journal of Molecular Sciences, 20*(21), 5491. https://doi.org/10.3390/ijms20215491

Patil-Gadhe, A., & Pokharkar, V. (2013). Single step spray drying method to develop proliposomes for inhalation: A systematic study based on quality by design approach. *Pulmonary Pharmacology & Therapeutics, 27*(2), 197–207. https://doi.org/10.1016/j.pupt.2013.07.006

Pokharkar, V., Patil-Gadhe, A., Kyadarkunte, A., Pereira, M., Jejurikar, G., Patole, M., & Risbud, A. (2014). Rifapentine-proliposomes for inhalation: In vitro and In vivo toxicity. *Toxicology International, 21*(3), 275. https://doi.org/10.4103/0971-6580.155361

Rojanarat, W., Changsan, N., Tawithong, E., Pinsuwan, S., Chan, H.-K., & Srichana, T. (2011). Isoniazid Proliposome Powders for Inhalation—Preparation, Characterization and Cell Culture Studies. *International Journal of Molecular Sciences, 12*(7), 4414–4434. https://doi.org/10.3390/ijms12074414

Rojanarat, W., Nakpheng, T., Thawithong, E., Yanyium, N., & Srichana, T. (2012a). Inhaled pyrazinamide proliposome for targeting alveolar macrophages. *Drug Delivery, 19*(7), 334–345. https://doi.org/10.3109/10717544.2012.721144

Rojanarat, W., Nakpheng, T., Thawithong, E., Yanyium, N., & Srichana, T. (2012b). Levofloxacin-Proliposomes: Opportunities for Use in Lung Tuberculosis. *Pharmaceutics, 4*(3), 385–412. https://doi.org/10.3390/pharmaceutics4030385

Singodia, D., Verma, A., Khare, P., Dube, A., Mitra, K., & Mishra, P. R. (2012). Investigations on feasibility of *in situ* development of amphotericin B liposomes for industrial applications. *Journal of Liposome Research, 22*(1), 8–17. https://doi.org/10.3109/08982104.2011.584317

Smith, I. (2003). *Mycobacterium tuberculosis* Pathogenesis and Molecular Determinants of Virulence. *Clinical Microbiology Reviews, 16*(3), 463–496. https://doi.org/10.1128/CMR.16.3.463-496.2003

Solleti, V. S., Alhariri, M., Halwani, M., & Omri, A. (2015). Antimicrobial properties of liposomal azithromycin for Pseudomonas infections in cystic fibrosis patients. *Journal of Antimicrobial Chemotherapy, 70*(3), 784–796. https://doi.org/10.1093/jac/dku452

Srichana, T., Eze, F. N., & Thawithong, E. (2022). A facile one-step jet-milling approach for the preparation of proliposomal dry powder for inhalation as effective delivery system for anti-TB therapeutics. *Drug Development and Industrial Pharmacy, 48*(10), 528–538. https://doi.org/10.1080/03639045.2022.2135101

Tongkanarak, K., Loupiac, C., Neiers, F., Chambin, O., & Srichana, T. (2024). Evaluating the biomolecular interaction between delamanid/formulations and human serum albumin by fluorescence, CD spectroscopy and SPR: Effects on protein conformation, kinetic and thermodynamic parameters. *Colloids and Surfaces B: Biointerfaces*, *239*, 113964. https://doi.org/10.1016/j.colsurfb.2024.113964

Vuong, N. N., Hammond, D., & Kontoyiannis, D. P. (2023). Clinical Uses of Inhaled Antifungals for Invasive Pulmonary Fungal Disease: Promises and Challenges. *Journal of Fungi*, *9*(4), 464. https://doi.org/10.3390/jof9040464

Wang, Q., Liu, J., Hu, Y., Pan, T., Xu, Y., Yu, J., Xiong, W., Zhou, Q., & Wang, Y. (2021). Local administration of liposomal-based Srpx2 gene therapy reverses pulmonary fibrosis by blockading fibroblast-to-myofibroblast transition. *Theranostics*, *11*(14), 7110–7125. https://doi.org/10.7150/thno.61085

Wang, C., Pan, C., Yong, H., Wang, F., Bo, T., Zhao, Y., Ma, B., He, W., & Li, M. (2023). Emerging non-viral vectors for gene delivery. *Journal of Nanobiotechnology*, *21*(1), 272. https://doi.org/10.1186/s12951-023-02044-5

Wang, J., Guo, Y., Lu, W., Liu, X., Zhang, J., Sun, J., & Chai, G. (2024). Dry powder inhalation containing muco-inert ciprofloxacin and colistin co-loaded liposomes for pulmonary *P. aeruginosa* biofilm eradication. *International Journal of Pharmaceutics*, *658*, 124208. https://doi.org/10.1016/j.ijpharm.2024.124208

Willis, L., Hayes, D., & Mansour, H. M. (2012). Therapeutic Liposomal Dry Powder Inhalation Aerosols for Targeted Lung Delivery. *Lung*, *190*(3), 251–262. https://doi.org/10.1007/s00408-011-9360-x

Wood, S. J., Kuzel, T. M., & Shafikhani, S. H. (2023). *Pseudomonas aeruginosa*: Infections, Animal Modeling, and Therapeutics. *Cells*, *12*(1), 199. https://doi.org/10.3390/cells12010199

Yan, L., Hou, C., Liu, J., Wang, Y., Zeng, C., Yu, J., Zhou, T., Zhou, Q., Duan, S., & Xiong, W. (2023). Local administration of liposomal-based Plekhf1 gene therapy attenuates pulmonary fibrosis by modulating macrophage polarization. *Science China Life Sciences*, *66*(11), 2571–2586. https://doi.org/10.1007/s11427-022-2314-8

Ye, T., Sun, S., Sugianto, T. D., Tang, P., Parumasivam, T., Chang, Y. K., Astudillo, A., Wang, S., & Chan, H.-K. (2018). Novel combination proliposomes containing tobramycin and clarithromycin effective against *Pseudomonas aeruginosa* biofilms. *International Journal of Pharmaceutics*, *552*(1–2), 130–138. https://doi.org/10.1016/j.ijpharm.2018.09.061

Zhang, Y., Shi, W., Zhang, W., & Mitchison, D. (2014). Mechanisms of Pyrazinamide Action and Resistance. *Microbiology Spectrum*, *2*(4). https://doi.org/10.1128/microbiolspec.MGM2-0023-2013

Zhang, M., Wang, Q., Wan, K.-W., Ahmed, W., Phoenix, D. A., Zhang, Z., Elrayess, M. A., Elhissi, A., & Sun, X. (2019). Liposome mediated-CYP1A1 gene silencing nanomedicine prepared using lipid film-coated proliposomes as a potential treatment strategy of lung cancer. *International Journal of Pharmaceutics*, *566*, 185–193. https://doi.org/10.1016/j.ijpharm.2019.04.078

Zou, Y., Zong, G., Ling, Y.-H., & Perez-Soler, R. (2000). Development of cationic liposome formulations for intratracheal gene therapy of early lung cancer. *Cancer Gene Therapy*, *7*(5), 683–696. https://doi.org/10.1038/sj.cgt.7700156

# Proliposomes for Pulmonary Drug Delivery: A Conclusive Summary

**5**

### Abstract

This chapter serves as a conclusive summary of the former four chapters regarding the field of proliposomes for pulmonary drug delivery. The growing prevalence of pulmonary diseases justifies the need for effective drug delivery directly to the lung for local treatment of pulmonary diseases, aiming to maximize therapeutic benefit and minimize systemic adverse effects. Liposomes have emerged as safe nanocarriers for inhalation, but their instability issues tend to slow down development of many new inhaled liposome formulations. Proliposomes are liposome precursor formulations, and they have emerged as alternatives to traditional liposome formulations, with potential for advancing the field of pulmonary inhalation of liposomes. Delivery of proliposomes and liposomes generated from proliposomes can be achieved using pMDIs, and more commonly DPIs and medical nebulizers. Characterization of proliposomes and liposomes is essential for formulation development, and subsequent investigations of proliposomal/liposomal aerosol revealed their suitability for "deep lung" deposition. Numerous studies conducted in our group and others have shown the advantage of using proliposome formulations as an approach for generating inhalable liposomes or proliposomes that may generate in situ within the lung. In vivo studies using experimental animals have revealed many possible therapeutic applications of liposomes generated from proliposomes including treatment of infectious diseases and cancer.

## 5.1 Introduction

There is a growing interest in pulmonary drug delivery owing to its non-invasive nature and the large surface area available for absorption (approximately $100 \text{ m}^2$). The prevalence of respiratory diseases, including asthma and chronic obstructive pulmonary disease

© The Author(s), under exclusive license to Springer Nature Switzerland AG 2025
A. Elhissi et al., *Proliposomes: A Manufacturing Technology of Liposomes for Pulmonary Drug Delivery*, Synthesis Lectures on Biomedical Engineering,
https://doi.org/10.1007/978-3-319-01297-1_5

(COPD), justifies the need to design effective formulations for pulmonary drug delivery (Rudokas et al. 2016).

For delivery of drugs to the lung, inhalation devices should be used. These are divided into three primary types: pressurized metered dose inhalers (pMDIs), dry powder inhalers (DPIs), and medical nebulizers (Laube and Dolovich 2014). Each of the three main types of inhalation devices is designed to fit particular types of formulations. Thus, while pMDIs are designed to deliver precise doses of a drug solubilized or dispersed in a liquefied propellant, DPI devices are designed to deliver drug dispersed or loaded into carrier particles that should eventually have appropriate aerodynamic properties to deposit into the "deep lung". By contrast, medical nebulizers are relatively bulky compared to pMDIs and DPIs, but can deliver large volumes (e.g. 2–5 ml) of drug solution or dispersion over a period of time (e.g. 10 min); thus, large drug doses are delivered to the lung. Medical nebulizers are divided into three types, namely, air-jet, ultrasonic and vibrating-mesh nebulizers (Laube and Dolovich 2014; Rudokas et al. 2016).

## 5.2 Challenges Associated with Pulmonary Drug Delivery

Drug delivery to the lung is challenged by a number of obstacles, including the difficulty to aerosolize the drug into "respirable" particles, which should be less than 5 μm to maximize the possibility of deposition in the "deep lung" (i.e. respiratory bronchioles and alveolar region) where the drug would be "therapeutically useful". For this to be achieved, appropriate formulations should be designed and effective inhalation devices need to be employed. Owing to the thinness of the pulmonary epithelium and abundance of blood supply in the lung tissue, the drug residence in the lung is short-lived; this necessitates frequent dosing which can cause increased potential of systemic adverse effects (D'Angelo et al. 2015).

## 5.3 Liposomes and Proliposomes

Liposomes are safe drug delivery systems, since they are manufactured using materials that are similar to mammalian biological membranes (e.g. phospholipids and cholesterol); thus, liposomes are biocompatible and biodegradable. Liposomes can solubilize water-insoluble drugs and entrap a wide range of drug molecules, offering controlled drug release in the lung upon inhalation, and potentially reducing systemic adverse effects. However, liposomes are unstable, chemically owing to possible hydrolysis and oxidation of the liposomal phospholipids, and microbiologically since they are similar to biological cells in composition. Freeze-drying has been introduced as a possible stabilizing step to liposomes since liposomes can be dried into "fluffy" solid product, under reduced pressure. However, freezing and negative pressure are both detrimental to the integrity of liposome structures; thus, freeze-drying may induce damage to liposomes, resulting in aggregation or fusion, with concomitant loss of the originally entrapped drug (Ball et al. 2016; Sainaga Jyothi et al. 2022).

Proliposome technologies have been introduced as an alternative formulation strategy of liposomes. Proliposomes serve as liposome precursors, which are either particulate-based (powders or granules) or solvent-based (commonly alcohol-based) (Payne et al. 1986; Perrett et al. 1991; Elhissi and Taylor 2005; Dhiman et al. 2022). Addition of aqueous phase causes spontaneous generation of liposomes that have characteristics dependent on excipients, manufacturing method and hydration procedure. Proliposomes are novel technologies that addresses challenges associated with traditional liposome formulations, such as storage, stability, and scalability for large-scale production. Particulate-based proliposomes can be manufactured on a small scale using a modified rotary evaporator with a feedline (Payne et al. 1986; Elhissi and Taylor 2005), or the slurry method within a rotary evaporator (Khan et al. 2015). However, large-scale manufacture is also possible using fluidized-bed coating (Gala et al. 2015), or spray-drying (Omer et al. 2018). The generated liposomes can be processed on a relatively large scale using high-pressure homogenization to generate small unilamellar liposomes (Beltrán et al. 2020).

## 5.4  Characterization Studies

Proliposomes should be characterized in terms of surface morphology using scanning electron microscopy (SEM). This is particularly important to predict whether the proliposomes would have good flowability for delivery from DPI devices. Stability and ability to enhance the drug dispersion properties can be investigated using differential scanning calorimetry and X-ray diffraction, respectively. For liposomes, it is important to analyze their size to ensure their suitability for delivery via nebulization. Furthermore, analysis of the particle size of aerosol incorporating liposomes/proliposomes is essential for the determination of fine particle fraction (FPF) and possibility of the drug to deposit in the 'deep lung'. Aerosol particle size is influenced by an interplay of factors including formulation, inhalation device design and delivery parameters. Drug encapsulation efficiency (EE) is a key measure of how effectively drugs are incorporated into liposome structures. A high EE is essential for achieving therapeutic goals (Khan et al. 2021).

## 5.5  Pulmonary Delivery of Proliposomes

The delivery of proliposomes from pMDIs has been achieved through dissolving the lipid components in a chlorofluorocarbon (CFC) propellant, while the drug is dissolved or dispersed in the pressurized solution depending on its physicochemical properties. In this approach, it is proposed that the delivered formulation will be hydrated to form liposomes in situ within the respiratory tract (Vyas et al. 2004). The validity of this assumption has been justified using an impinger containing aqueous phase. Due to the harmful effect of CFC propellants on the ozone layer, these have been replaced by hydrofluoroalkane (HFA) propellants, which are not good at solubilizing phospholipids, resulted in limited use of pMDIs for the delivery of proliposomes and liposomes. The delivery of proliposomes using DPIs is more promising and many successful achievements have been published, for example using spray drying (Rojanarat et al. 2012; Omer et al. 2018). So far, the most successful inhalation device for delivering traditional liposomes and vesicles generated from proliposomes are nebulizers, with air-jet and vibrating-mesh nebulizers demonstrating more promising findings compared to ultrasonic nebulizers (Elhissi and Taylor 2005). Many studies demonstrated the success of nebulizers at delivering liposomes generated from proliposomes. Arikace is the first and only FDA liposomal formulation approved for inhalation (Bilton et al. 2021). It is possible and logical that attempts in the future will use proliposome technologies to generate liposomes that are eligible for FDA approval.

## 5.6    Potential Therapeutic Applications of Proliposome Formulations

Research investigations have reported that proliposomes can facilitate both local and systemic drug delivery to the lungs, potentially allowing for targeted treatment of asthma, lung cancer, cystic fibrosis, and pulmonary infections. Various therapeutic applications of proliposomes have been researched, particularly in treating respiratory tract infections such as tuberculosis (TB). TB is notably challenging to treat due to the ability of *Mycobacterium tuberculosis* to thrive within macrophages and develop antimicrobial resistance. The use of proliposomal formulations has shown promise in overcoming these challenges. For example, studies have reported successful designs of proliposomal formulations encapsulating antitubercular drugs like rifampicin and isoniazid, demonstrating improved aerosolization properties, sustained drug release, and enhanced efficacy against TB (Parhizkar et al. 2021). In addition to TB, the application of proliposomes in fighting against *Pseudomonas aeruginosa* infections, particularly in cystic fibrosis patients has been investigated. Proliposomal formulations containing aminoglycosides have been explored and shown to enhance antimicrobial activity of the incorporated drug. Studies indicated that formulations such as tobramycin and clarithromycin combinations in proliposome formulations could demonstrate synergistic effects against *P. aeruginosa* biofilms, highlighting the great potential of proliposomes (Ye et al. 2018).

The investigations on proliposomes extends to fungal infections, with amphotericin B being a model antifungal agent. Aerosolized antifungal treatments prepared using proliposomes have shown promising results in targeting pulmonary fungal infections. Studies indicated that inhalable proliposomal formulations of amphotericin B might facilitate targeted delivery and improve therapeutic outcomes while minimizing systemic adverse effects of the drug (Albasarah et al. 2010).

Asthma is another disease that proliposomes have been used for incorporating anti-asthma bronchodilators and steroids. Proliposomes encapsulating the bronchodilator salbutamol sulphate have been developed and analyzed for their ability to enhance pulmonary delivery. Research indicates that the choice of carrier and formulation techniques significantly influence the performance of proliposome formulations (Omer et al. 2018).

Lung cancer treatment is another area of study. Proliposomes offer a strategy for direct delivery of chemotherapeutic agents to lung tissues. In vitro studies involving paclitaxel-loaded proliposomes suggested that formulations are more toxic to cancer cells compared to healthy tissues (Khan et al. 2020). The document also highlights the incorporation of investigational therapies, such as curcumin in proliposomal formulations, which have shown enhanced lung penetration and improved pharmacokinetics using experimental animals (Adel et al. 2021).

Finally, proliposomes have been investigated as gene carriers. Gene therapy provides an innovative approach for managing various genetic disorders affecting pulmonary function. The use of liposomes as nonviral vectors for gene delivery has shown promise in enhancing transfection efficiency and reducing adverse effects. Research on cationic liposomes delivering siRNA for gene silencing in lung cancer models illustrates the potential of proliposomal formulations for targeted gene therapy using experimental animal models (Zhang et al. 2019).

## 5.7    Final Conclusions

Liposomes are established delivery systems for inhalation, evidenced by the FDA approval of the liposomal Amikacin Arikayce® for the treatment of *Mycobacterium avium* complex (MAC) pulmonary infection via inhalation using the Pari *e-Flow* vibrating-mesh nebulizer (Ferreira et al. 2021). With the recent advancements in the use of liposomes, not only through the pulmonary drug delivery route, but also through other routes of administration including oral, transdermal and parenteral, we expect more liposomal products to be clinically approved in the future.

Research has clearly demonstrated that proliposome technologies provide feasibility and convenience for the preparation of liposomes in various size ranges for the incorporation of various therapeutic molecules. The formulation of proliposomes relied on widely used strategies in pharmaceutical industry such as fluid-bed coating, spray drying and high-pressure homogenization. Proliposomes also involve very safe excipients such as carbohydrates like sucrose, lactose, sorbitol and mannitol, which are established food constituents and clinically approved pharmaceutical excipients. Furthermore, like liposomes made by thin-film hydration (i.e. traditional liposomes), vesicles generated using proliposome systems are made from phospholipids and cholesterol, which are the main compositions of mammalian biological membranes. Solvent-based proliposome systems involve the use of very small concentrations of ethanol, which is an established co-solvent in pharmaceutical formulations including those used clinically and cosmetically.

Many important studies have been conducted on liposomes generated from proliposomes including their use for antimicrobial, anticancer and antiasthma applications. These studies demonstrated both safety and efficacy of these formulations, not only in vitro, but also in vivo using experimental animal models. The aerosol characteristics of proliposomes (as DPIs), or proliposome-generated liposomes via nebulization demonstrate the capability of these formulations to deposit in the "deep lung", with evidenced delivery of high dose fractions of the drug. Thus, we expect that more pulmonary drug delivery research on proliposomes will focus on further in vivo studies on animals with subsequent clinical studies.

In conclusion, the advantages of proliposomal formulations in pulmonary drug delivery demonstrate their potential in treating many conditions including pulmonary infections, asthma, and cancer. The stability, biocompatibility, and versatility of proliposomes represent significant advancements in drug delivery systems, paving the way for improved therapeutic outcomes. Further in vivo research is needed to establish the clinical applicability of proliposomes for pulmonary drug delivery.

## References

Adel, I. M., ElMeligy, M. F., Abdelrahim, M. E., Maged, A., Abdelkhalek, A. A., Abdelmoteleb, A. M., & Elkasabgy, N. A. (2021). Design and Characterization of Spray-Dried Proliposomes for the Pulmonary Delivery of Curcumin. International Journal of Nanomedicine, Volume 16, 2667–2687. https://doi.org/10.2147/IJN.S306831

Albasarah, Y. Y., Somavarapu, S., Stapleton, P., & Taylor, K. M. G. (2010). Chitosan-coated antifungal formulations for nebulisation. Journal of Pharmacy and Pharmacology, 62(7), 821–828. https://doi.org/10.1211/jpp.62.05.0002

Ball, R., Bajaj, P., & Whitehead, K. (2016). Achieving long-term stability of lipid nanoparticles: examining the effect of pH, temperature, and lyophilization. International Journal of Nanomedicine, Volume 12, 305–315. https://doi.org/10.2147/IJN.S12306237

Beltrán, J. D., Ricaurte, L., Estrada, K. B., & Quintanilla-Carvajal, M. X. (2020). Effect of homogenization methods on the physical stability of nutrition grade nanoliposomes used for encapsulating high oleic palm oil. LWT, 118. https://doi.org/10.1016/j.lwt.2019.108801

Bilton, D., Fajac, I., Pressler, T., Clancy, J. P., Sands, D., Minic, P., Cipolli, M., Galeva, I., Solé, A., Quittner, A. L., Jumadilova, Z., Ciesielska, M., & Konstan, M. W. (2021). Long-term amikacin liposome inhalation suspension in cystic fibrosis patients with chronic *P. aeruginosa* infection. Journal of Cystic Fibrosis, 20(6), 1010–1017. https://doi.org/10.1016/j.jcf.2021.05.013

D'Angelo, I., Conte, C., Miro, A., Quaglia, F., & Ungaro, F. (2015). Pulmonary drug delivery: a role for polymeric nanoparticles? *Current Topics in Medicinal Chemistry*, 15, 386–400

Dhiman, N., Sarvaiya, J., & Mohindroo, P. (2022). A drift on liposomes to proliposomes: recent advances and promising approaches. Journal of Liposome Research, 32(4), 317–331. https://doi.org/10.1080/08982104.2021.2019762

Elhissi, A. M. A., & Taylor, K. M. G. (2005). Delivery of liposomes generated from proliposomes using air-jet, ultrasonic, and vibrating-mesh nebulisers. Journal of Drug Delivery Science and Technology, 15(4), 261–265. https://doi.org/10.1016/S1773-2247(05)50047-9

Ferreira, M., Ogren, M., Dias, J. N. R., Silva, M., Gil, S., Tavares, L., Aires-da-Silva, F., Gaspar, M. M., & Aguiar, S. I. (2021). Liposomes as Antibiotic Delivery Systems: A Promising Nanotechnological Strategy against Antimicrobial Resistance. Molecules, 26(7), 2047. https://doi.org/10.3390/molecules26072047

Gala, R. P., Khan, I., Elhissi, A. M. A., & Alhnan, M. A. (2015). A comprehensive production method of self-cryoprotected nano-liposome powders. International Journal of Pharmaceutics, 486(1–2), 153–158. https://doi.org/10.1016/j.ijpharm.2015.03.038

Khan, I., Yousaf, S., Subramanian, S., Korale, O., Alhnan, M. A., Ahmed, W., Taylor, K. M. G., & Elhissi, A. (2015). Proliposome powders prepared using a slurry method for the generation of beclometasone dipropionate liposomes. International Journal of Pharmaceutics, 496(2), 342–350. https://doi.org/10.1016/j.ijpharm.2015.10.002

Khan, I., Lau, K., Bnyan, R., Houacine, C., Roberts, M., Isreb, A., Elhissi, A., & Yousaf, S. (2020). A Facile and Novel Approach to Manufacture Paclitaxel-Loaded Proliposome Tablet Formulations of Micro or Nano Vesicles for Nebulization. Pharmaceutical Research, 37(6), 116. https://doi.org/10.1007/s11095-020-02840-w

Khan, I., Yousaf, S., Najlah, M., Ahmed, W., & Elhissi, A. (2021). Proliposome powder or tablets for generating inhalable liposomes using a medical nebulizer. Journal of Pharmaceutical Investigation, 51(1), 61–73. https://doi.org/10.1007/s40005-020-00495-8

Laube, B. L., & Dolovich, M. B. (2014). 66 - Aerosols and Aerosol Drug Delivery Systems. In Adkinson, N. F., Bochner, B. S., Burks, A. W., Busse, W. W., Holgate, S. T., Lemanske, R. F. & O'Hehir, R. E. (Eds.) Middleton's Allergy (Eighth Edition). London: W.B. Saunders

Omer, H. K., Hussein, N. R., Ferraz, A., Najlah, M., Ahmed, W., Taylor, K. M. G., & Elhissi, A. M. A. (2018). Spray-Dried Proliposome Microparticles for High-Performance Aerosol Delivery Using a Monodose Powder Inhaler. AAPS PharmSciTech, 19(5), 2434–2448. https://doi.org/10.1208/s12249-018-1058-4

Parhizkar, E., Sadeghinia, D., Hamishehkar, H., Yaqoubi, S., Nokhodchi, A., & Alipour, S. (2021). Carrier Effect in Development of Rifampin Loaded Proliposome for Pulmonary Delivery: A Quality by Design Study. Advanced Pharmaceutical Bulletin. https://doi.org/10.34172/apb.2022.032

Payne, N. I., Browning, I., & Hynes, C. A. (1986). Characterization of Proliposomes. Journal of Pharmaceutical Sciences, 75(4), 330–333. https://doi.org/10.1002/jps.2600750403

Perrett, S., Golding, M., & Williams, W. P. (1991). A Simple Method for the Preparation of Liposomes for Pharmaceutical Applications: Characterization of the Liposomes. Journal of Pharmacy and Pharmacology, 43(3), 154–161

Rojanarat, W., Nakpheng, T., Thawithong, E., Yanyium, N., & Srichana, T. (2012). Inhaled pyrazinamide proliposome for targeting alveolar macrophages. Drug Delivery, 19(7), 334–345. https://doi.org/10.3109/10717544.2012.721144

Rudokas, M., Najlah, M., Alhnan, M. A., & Elhissi, A. (2016). Liposome Delivery Systems for Inhalation: A Critical Review Highlighting Formulation Issues and Anticancer Applications. Medical Principles and Practice, 25(Suppl. 2), 60–72. https://doi.org/10.1159/000445116

Sainaga Jyothi, V. G. S., Bulusu, R., Venkata Krishna Rao, B., Pranothi, M., Banda, S., Kumar Bolla, P., & Kommineni, N. (2022). Stability characterization for pharmaceutical liposome product development with focus on regulatory considerations: An update. International Journal of Pharmaceutics, 624, 122022. https://doi.org/10.1016/j.ijpharm.2022.12202237

Vyas, S. P., Kannan, M. E., Jain, S., Mishra, V., & Singh, P. (2004). Design of liposomal aerosols for improved delivery of rifampicin to alveolar macrophages. International Journal of Pharmaceutics, 269(1), 37–49. https://doi.org/10.1016/j.ijpharm.2003.08.017

Ye, T., Sun, S., Sugianto, T. D., Tang, P., Parumasivam, T., Chang, Y. K., Astudillo, A., Wang, S., & Chan, H.-K. (2018). Novel combination proliposomes containing tobramycin and clarithromycin effective against *Pseudomonas aeruginosa* biofilms. International Journal of Pharmaceutics, 552(1–2), 130–138. https://doi.org/10.1016/j.ijpharm.2018.09.061

Zhang, M., Wang, Q., Wan, K.-W., Ahmed, W., Phoenix, D. A., Zhang, Z., Elrayess, M. A., Elhissi, A., & Sun, X. (2019). Liposome mediated-CYP1A1 gene silencing nanomedicine prepared using lipid film-coated proliposomes as a potential treatment strategy of lung cancer. International Journal of Pharmaceutics, 566, 185–193. https://doi.org/10.1016/j.ijpharm.2019.04.078